D0458453

WITHDRAWN

The End of the Dinosaurs
Chicxulub Crater and Mass Extinctions

The End of the Dinosaurs gives a detailed account of the great mass extinction that rocked the Earth 65 million years ago, and focuses on the discovery of the culprit: the Chicxulub impact crater in Mexico. It recounts the birth of the cosmic hypothesis, the controversy that preceded its acceptance, the search for the crater, its discovery and ongoing exploration, and the effect of the giant impact on the biosphere. Other mass extinctions in the fossil record are reviewed, as is the threat of asteroids and comets to our planet today. The account of the impact and its aftermath is suitable for general readers. The description of the crater geology is given in enough detail to interest students of the Earth sciences. A detailed index and bibliography are included.

CHARLES FRANKEL studied planetary geology at Middlebury College, Vermont, and the University of Arizona. He has directed educational and documentary films on geology, astronomy and space exploration. Mr Frankel lives in Los Angeles and Paris.

CHARLES FRANKEL

The End of the Dinosaurs

Chicxulub Crater and Mass Extinctions

CAMBRIDGE
UNIVERSITY PRESS

PUBLISHED BY THE PRESS SYNDICATE OF THE UNIVERSITY OF CAMBRIDGE
The Pitt Building, Trumpington Street, Cambridge, United Kingdom

CAMBRIDGE UNIVERSITY PRESS
The Edinburgh Building, Cambridge CB2 2RU, UK www.cup.cam.ac.uk
40 West 20th Street, New York, NY 10011-4211, USA www.cup.org
10 Stamford Road, Oakleigh, Melbourne 3166, Australia
Ruiz de Alarcón 13, 28014 Madrid, Spain

Original edition published in French as *La mort des dinosaures*
by Charles Frankel

First published 1999

Printed in the United Kingdom at the University Press, Cambridge

Typeset in Postscript Ehrhardt 11/13pt [RO]

A catalogue record for this book is available from the British Library

Library of Congress Cataloguing in Publication data

Frankel, Charles.
[Mort des dinosaures. English]
The end of the dinosaurs: Chicxulub crater and mass extinctions/Charles Frankel.
p. cm.
Includes bibliographical references and index.
ISBN 0 521 47447 7 (hardbound)
1. Catastrophes (Geology) 2. Extinction (Biology)
3. Cryptoexplosion structures – Mexico – Campeche, Bay of, Region.
4. Chicxulub Crater. I. Title.
QE506.F7313 1999
551.3'97 – dc21 98-49427 CIP

ISBN 0 521 47447 7 hardback

Contents

Introduction

This book tells of the discovery of Chicxulub impact crater in Mexico, and why this 180 kilometer-wide collision scar is held responsible for the massacre of countless species at the end of the Cretaceous: the great 'K-T' mass extinction.

The thriving dinosaur community was the most celebrated victim of the extinction, making it a household mystery shared by scientists, the media and the public alike.

Not only does the discovery of Chicxulub crater bring a convincing answer to the long-debated mystery of the mass extinction: it also stands as a brilliant example of the scientific method at work. The trail to the crater was uncovered through clever induction, testing and validation of bold hypotheses. Our goal in this book is to retrace this exciting string of discoveries that led scientists from the hills of Umbria in Italy to the cuestas of Colorado and finally to the Gulf of Mexico and the Yucatan – an adventure marked by brilliant intuitions, painstaking fieldwork, and a jungle of false leads and controversies.

In telling the story of Chicxulub crater and the end of the dinosaurs, it is hard to remain objective. Seldom has a scientific discovery been so controversial and so highly profiled in the media. Like everyone, I have my own preferences where different models are proposed and different interpretations possible. My intent is not to present a definitive historical account of K-T research and discoveries, but to tell as entertaining a story as possible, based on my own appraisal of the scientific literature and on discussions with several of the major players in the field. In following my storyline, I leave out the work of countless researchers: may they forgive me for not quoting their work! As for those I do quote, I hope I have not deformed or oversimplified their contributions.

The study of mass extinction and the end of the dinosaurs cuts across a variety of fields, such as paleontology (the study of fossils and ancient lifeforms), biology and evolution, planetary geology and astronomy, geophysics and geochemistry, climatology and environmental science. It is this overlapping of so many different fields that makes K–T research so exciting. Conversely, there are as many different approaches to the problem as there are disciplines. Hence it is only fair that I state my own background, which influenced how I covered the issue in this book.

My training is in planetary geology, at Middlebury College (Vermont) and the University of Arizona. The discovery of the Chicxulub impact crater caught my attention in 1992 and led me to assess its ties with the extinction of the dinosaurs, rather than the other way around: most researchers worked their way instead from the extinction of the dinosaurs to the discovery of the crater!

This 'after-the-fact' interest in the demise of the dinosaurs gave me some perspective to assess the raging controversy of their extinction, but does not endow me with any more objectivity than my fellow researchers: my biases and convictions remain.

I have divided this book into eight chapters, and for each one asked a specialist to review its content. I am greatly indebted to those scientists who took the time to correct my mistakes and suggest improvements, especially since they did not always agree with my coverage of the issues. All errors that remain are mine, particularly since my reviewers did not always have a chance to check my latest revisions. Hereunder I briefly describe the thrust of each chapter and take the opportunity to thank my reviewers.

Chapter one sets the stage of the great controversy surrounding the end of the dinosaurs, which led to the realization that this disappearance was sudden, rather than progressive. To read over this first chapter, no one was more qualified than Canadian paleontologist Dale Russell, now at the University of North Carolina, who was one of the first experts on dinosaurs to recognize their brutal and cosmic demise.

Chapter two describes the chain of evidence that established the impact theory, from the landmark article by Alvarez *et al.* in 1980 to the discovery of shocked minerals and meteoritic material confirming the hypothesis. I asked French geochemists Robert Rocchia and Eric Robin, from the CEA/CNRS laboratory, to review this chapter, since they were closely involved in assessing the evidence in a country that was especially skeptical with regard to the impact theory.

This skepticism is the focus of the third chapter, which tells of the mounting controversy in the 1980s, and the alternative theories construed to explain away the impact evidence. Dale Russell and Robert Rocchia were kind enough to review this chapter as well: they were both close observers of the raging debate in their respective countries.

Chapter four describes the search for the impact crater, and what planetary geologists knew of such impact scars as the search effort grew in the late 1980s. Impact specialist Richard A.F. Grieve, of Natural Resources Canada, took time to review this chapter and contribute several illustrations, including the up-to-date map of terrestrial impact craters.

Chapter five tells of the discovery of the buried Chicxulub crater in Yucatan and the ongoing surveys that strive to constrain its shape and size. I am grateful to Alan R. Hildebrand of Natural Resources Canada, one of the original discoverers of the crater, for having reviewed this important chapter, and for contributing many illustrations as well. He was an inspiration throughout the writing of this book.

Chapter six describes the aftermath of the impact and the lethal mechanisms that rocked the biosphere. Specialist Owen B. Toon of NASA's Ames Research Center provided me with much insight into the various scenarios that are presently being evaluated.

In chapter seven, we broaden the issue by examining the other mass extinctions of the fossil record and searching for evidence of impact at those boundaries as well. Michael Rampino of New York University, who was one of the first scientists to generalize the impact theory to all mass extinctions, reviewed this section.

Chapter eight concludes this book by reviewing the impact hazard here and now, on the timescale of human civilization, and explores what steps can be taken to avoid an asteroid or comet hit in the future. Asteroid specialist Tom Gehrels of the University of Arizona, director of the Spacewatch detection program, advised me on this final chapter.

A bibliography lists a few of the landmark books on impacts and mass extinctions, as well as a hundred or so research articles – a very limited sample of the literature since over 2000 articles have been published on the K-T extinction alone!

I also wish to thank all the institutions and individuals that have contributed illustrations to this book, and especially William K. Hartmann for his wonderful artwork. I extend my deep thanks as well to the staff of Masson Press in Paris for publishing the original version of this book in French. My hope is now that this fully revised and updated edition will live up to its subject and convey some of the excitement that has animated paleontology and planetary science over the last two decades.

Paris

1
The great mass extinctions

The end of the dinosaurs has been a challenge to scientists since the founding days of geology. A once dominant group of animals of tremendous vigour and diversity, their unexplained demise at the end of the Cretaceous was a challenge to the very theory of evolution and the law of the fittest. Many models were proposed to account for the mysterious fall from grace of the dinosaurs, and this book will focus on the one thesis that is now supported by the evidence, and has forever changed our vision of life on Earth: the thesis of a cataclysmic cosmic impact that raised havoc on planet Earth.

Before we follow the string of evidence that leads from the last fossils of the Cretaceous to the Chicxulub impact crater in Yucatan – the focal point of this book – let us first expose the background to our story, and notably where the question of the extinction of the dinosaurs stands in the adventure of science.

A challenge to scientists

The story begins with paleontology, the science of fossils. As they went about collecting and classifying fossils, the founders of paleontology discovered that there was something odd in the time succession of past lifeforms on Earth. There were puzzling levels in the sedimentary record – representing only 'moments' in geological time – where a number of fossil types ended abruptly, and were replaced above the boundary level by new groups of burgeoning species. What, pondered the scientists, could possibly be the cause of such dramatic turnovers of plant and animal life?

French nobleman Georges Cuvier (1769–1832), one of the founding fathers of paleontology, ventured that these sharp transitions were the sign of brutal, planet-wide catastrophes.[1] After each such 'revolution', Cuvier claimed, exotic, little-known species would move in from the regions that had been spared, and would eventually repopulate the Earth.

In support of 'catastrophism', Cuvier and his disciples observed that the 'revolution' boundaries in the rock strata were generally marked by abrupt changes in sedimentation style – notably layers of chaotic rock, as if catastrophes had effectively taken place.

There was an opposing school of thought that staunchly rejected these catastrophic views. Englishman Charles Lyell, one of the most influential geologists of his time, was convinced of the smooth, progressive nature of geological processes, and dismissed the need for catastrophes. In his view and that of his followers, the Earth was a slowly evolving world, shaped by the ant-paced toil of erosion and sedimentation. So-called 'catastrophic' boundaries were only gaps in the sedimentary record, when long periods of time had not been recorded, or were erased by erosion, and thus gave the impression of a radical change across the line. In the 'gradualistic' view, the turnover of living species *had* to be slow, dictated by the laws of competition and natural selection – the theory founded in the mid-nineteenth century by Charles Darwin and Alfred Wallace.

Between these two schools of thought – gradualism and catastrophism – the scientific establishment came to favor Lyell's gradualistic approach, although a few undisputable breaks in the fossil record remained unexplained, and did split up the history of the Earth into markedly different periods, populated by distinct groups of plant and animal species (see figure 1.1).

[1] Baron Georges Cuvier was especially impressed by a major faunal turnover in the Paris Basin, occurring at the top of the chalk layers of the Cretaceous period – the End-Cretaceous 'K-T' boundary.

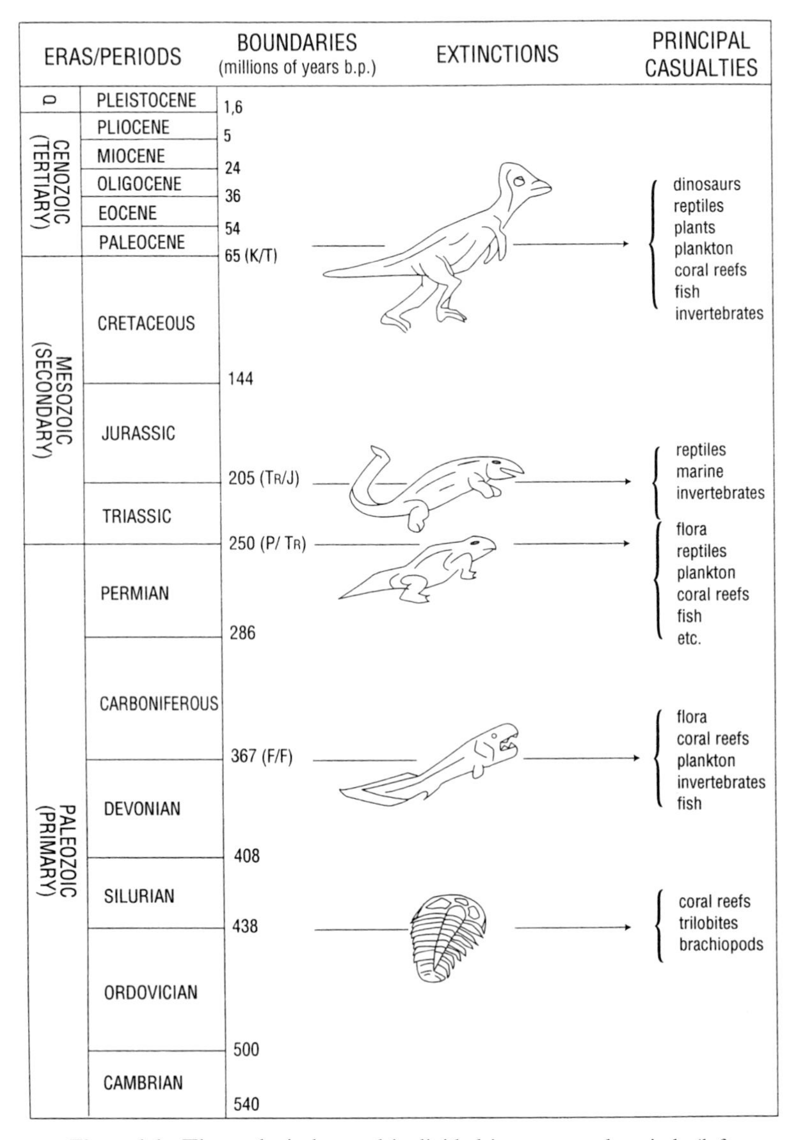

Figure 1.1. The geological record is divided into eras and periods (left columns), according to the fossils identified in the rock strata. Five of the boundaries between periods are marked by abrupt faunal turnovers, when most species on land and in the sea were exterminated. These are known as the great mass extinctions (the principal victims are listed in the right column).

The meaning of extinction

Before we examine the fossil record in more detail, let us define what we mean by a species. A species is a set of lifeforms, where all members of the group display similar characteristics, and within which reproduction takes place, the offspring carrying on the lineage. The characteristics of the species are stored in its DNA make-up, which is transmitted from generation to generation through the reproduction process.[2]

It is estimated that there are *thirty to forty million* different species living on Earth today, covering the whole gamut of viruses and bacteria, mushrooms, plants, insects, and marine, terrestrial and flying invertebrates and vertebrates, including the single human species. Biologists have studied and described in detail only one million species so far.

Other periods, in the past, have been equally well stocked. Since the blossoming of complex life in the Cambrian – roughly 600 million years ago – the number and variety of species have undergone a constant turnover, with new species appearing as others die out.

Species are indeed mortal: they are pronounced dead when all their members have vanished. The extinction of a species usually occurs over many generations, as a result of changing environmental conditions or increased competition from other species, which progressively cut down its numbers until reproduction is no longer possible within the group.

From our monitoring of the biosphere today, it appears that several species go extinct each year – that year seeing the last of their members die out. On the other hand, each year also sees the appearance of new species, branching out from other species to develop in new directions, notably under the influence of genetic mutations. All in all, the biosphere is then in a constant state of flux, with a small 'background' rate of extinctions balanced by an about equal birth rate of new species.

[2] Some amount of 'detail remodeling' takes place in the DNA through random mutations; favorable changes are amplified through the filter of natural selection.

The great mass extinctions

Above this background rate of appearances and extinctions in the fossil record stand out those unusual episodes where a great many extinctions appear to occur simultaneously – the so-called 'mass extinctions' that show up as conspicuous boundaries in the fossil record. Over half of the known species of a period can vanish in the largest of these global events, which number five in the geological record. These are referred to as the great mass extinctions, or 'the Big Five'.[3]

The earliest great mass extinction occurred at the end of the Ordovician period, 440 million years ago. At the time, life was confined to the oceans and had barely set its foothold on dry land. In the marine realm, numerous species of trilobites, plankton and coral disappeared over a narrow timespan – less than 500 000 years according to the latest findings.

Equally impressive was the following extinction at the close of the Devonian period, around 365 million years ago, in what appeared to be a succession of crises that hit plankton families and coral reef systems, trilobites, brachiopods and primitive fish in several waves.

This second mass extinction was followed by a long period of recovery and expansion of the marine and terrestrial biosphere (the Carboniferous and Permian periods), which saw the blossoming of the first branches of amphibian and reptile species, until a new, even more powerful blow struck the planet 250 million years ago.

This third great mass extinction, at the end of the Permian, saw the demise of 90 % of marine species, and the extinction of most terrestrial species (including the vast majority of the newly-established amphibians and reptiles). The timeframe of the crisis is still open to discussion. A few paleontologists believe that the biosphere declined in a progressive fashion over an interval of one to two million years, but recent work has convinced many researchers that the Permian mass extinction consisted of one or several short crises lasting tens of thousands of years at most.

[3] Close to twenty lesser mass extinctions have been identified in the fossil record, where the level of species killing exceeds 20 %.

After this lethal End-Permian extinction, the biosphere slowly recovered to see the rise of reptiles and dinosaurs. This new era in the history of life had hardly begun when it was stunted by the fourth great mass extinction, about 205 million years ago. In this End-Triassic crisis, the marine realm was the most severely affected, but on land as well there were many victims among reptiles and early dinosaurs.

Then followed the Jurassic and Cretaceous periods, which saw the blossoming of yet more new species, and the rise and reign of the dinosaurs. Then, 65 million years ago, the fifth great mass extinction struck Earth, wiping out an estimated 70 % of all species worldwide. In the oceans, plankton species were massacred, along with scores of species of bivalves, belemnites, spiral-shelled ammonites and other mollusks, fish, and great marine lizards, including the giant mosasaurs – swimming lizards several meters long and many tons in weight. On dry land, numerous families of reptiles and mammals were exterminated, as well as the totality of species of dinosaurs and flying reptiles.

Because it was spectacular, apparently short-lived, and closest to us in time, the crisis at the end of the Cretaceous is the best studied of the great mass extinctions, with a metaphysical twist that concerns us directly. Indeed, it can be argued that our mammalian ancestors took over the Earth in the wake of the drama, taking advantage of the ecological niches left vacant by the dinosaurs, and setting the stage for the ascent of man.[4]

The End-Cretaceous

Before we examine in detail this crucial End-Cretaceous mass extinction, and the detective work that led to the resolution of its mystery, let us brush a rough portrait of planet Earth in the final days that led up to the crisis.

Some 65 million years ago, the world looked somewhat different

[4] Some lifeforms have morphologies favored by evolution that allow them to survive major crises, for example fish and birds. The human species itself may be resistant to catastrophes: time will tell.

than it does today. For one, its geography was resolutely distinct, due to the different layout of tectonic plates. Continents and ocean basins were arranged differently at the time. Indeed, at the rate at which ocean-floor spreading takes place – several centimeters per year – playing the movie 65 million years back in time changes the relative position of most continents by several thousand kilometers (see figure 1.2).

In the End-Cretaceous, the continents were separated by a somewhat larger Pacific ocean but a narrower Atlantic basin. Ocean waters flowed freely around the globe at the Equator, through submerged sections of Central America between the Pacific and the Atlantic, and through an ancestral version of the Mediterranean – the Tethys sea – between the East Atlantic and the Indian Ocean. Freshly split off Africa, India was an island creeping northward towards Asia on the back of a tectonic plate, on its way to colliding with the continent and uplifting the Himalayan range. Further south, continental rifting was separating Australia from Antarctica.

One consequence of the different arrangement of ocean basins and rifts back in the Cretaceous was a higher sea level. Voluminous submarine relief displaced the water upwards onto the continental margins, flooding the lowlands. This marine high stand was all the more extreme in that there were no polar ice caps in the Late Cretaceous to pump water out of the ocean basins. Today, thousands of meters of ice are piled up in the polar caps of Antarctica and Greenland, and withdraw tens of meters of water from the oceans. But in the Late Cretaceous, the continents were too far away from the poles to freeze water precipitation, and the sea level could not be lowered.

Large underwater relief and lack of polar caps combined to put the sea level one hundred meters higher than today's mark, causing the seas to penetrate deeply into the plains and basins of Africa, South America, India, and especially North America and Europe. In North America, a vast inner sea penetrated from the Gulf of Mexico northward through the great plains up to Canada and the Arctic. As for Europe, its lowlands were flooded as far east as Russia and the Ural foothills. In this European shallow sea, highlands emerged as a balmy archipelago strung out from Iberia to Central

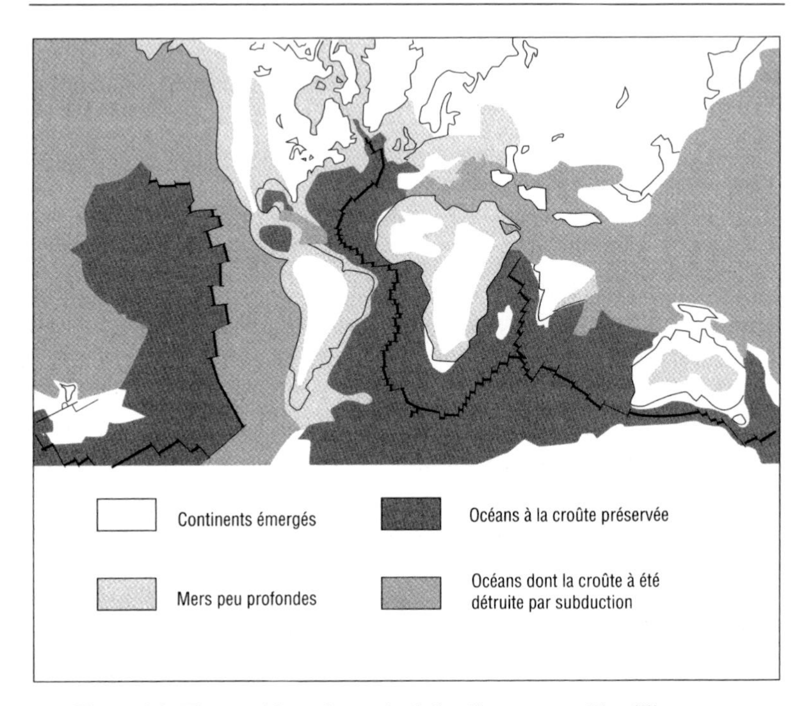

Figure 1.2 The world at the end of the Cretaceous, 65 million years ago. Continents above sea level are shown in white, continental margins and plains covered by shallow seas in light gray. Oceanic crust that has been destroyed by subduction since the End-Cretaceous is colored gray, and oceanic crust still present at the Earth's surface today is shown in black. 'Zipper' black lines indicate active submarine rifts. At the end of the Cretaceous, sea-floor spreading is established in the Atlantic basin, separating the Americas from Africa and Europe. Africa is about to drift north and close the ancestral Mediterranean (Tethys sea). India is creeping northward, on its way to colliding with Asia. (*Modified from a map by George Retsek, in Un impact d'origine extraterrestre, Walter Alvarez and Frank Asaro,* Pour la Science *(French edition of* Scientific American*), 1993.*)

France, northern Europe and Transylvania. To the south-east, beyond a deep sedimentary basin (which would later fold and rise to form the Alps), hillcrests and mountain blocks emerged as yet more islands, strung along the proto-Mediterranean sea.

Temperatures were higher in the Late Cretaceous than at

present. The flooded continents enjoyed milder climates, especially Europe, which was closer to the Equator. Coral reefs blossomed in the warm waters, along with other thriving marine communities. The remains of their shells form the extensive chalk beds and cliffs that are typical of the period.[5]

All in all, living species enjoyed widespread diversification in the Late Cretaceous as their continental 'rafts' moved away from each other under the impulse of plate tectonics, and rising sea levels fragmented their continental habitat. Dinosaurs were no exception. Field work by Paul Sereno and his colleagues in the Sahara pointed to a rapid diversification of the dinosaurian fauna in Africa at the time.[6] Provincial evolution of the dinosaurs was also well under way in North and South America, the European archipelago, India, and the Australia–Antarctica block.

With this global blooming of lifeforms in the Late Cretaceous came further change. The sea level had reached its maximum stand on the continents and was starting to decline again, probably because of a now deepening trend in the ocean basins that was draining back the marine waters.[7] As the sea level began to fall again and more continental area was exposed, the climate hardened in many parts of the globe. Different regions and biotopes were affected in different ways.

Paleontologists who study various ecosystems and fossil species around the world each have their own reading of the environmental conditions that prevailed at the end of the Cretaceous, before the mass extinction. Dinosaur experts in particular find it difficult to come up with a detailed picture because there are few dinosaur fossil sites of Late Cretaceous age. One string of precious sites extends across North America, from Texas to Alberta, along the shores of the inner continental sea: we shall take a look at these rich fossil beds in a later chapter. Hereunder we briefly visit those

[5] The name 'Cretaceous' comes from the Latin word for chalk: 'creta'. The symbol for this geological period is the letter 'K' (from the German word for chalk: 'Kreide').

[6] Serenos, P.C. *et al.*, Predatory dinosaurs from the Sahara and Late Cretaceous faunal differentiation, *Science*, **272**, 986–91, 1996.

[7] One reason for these 'transgressions' (sea level advances) and 'regressions' (sea level retreats) has to do with changes of geometry and pace in the puzzle of plate tectonics, with varying underwater relief causing variable water levels.

of Provence in southern France to introduce the variety of dinosaur species that roamed the Earth at the end of the Cretaceous, and to give but one example of the scores of hypotheses that were drummed up to explain their extinction, before the advent of the cosmic hypothesis.

Dinosaurs of Provence

At the foot of the Montagne Sainte-Victoire – the colorful hill that graces many a Cézanne painting – balmy swamps were teeming with life in the final days of the Cretaceous. For dinosaurs the Provence swamps must have been a mating paradise, if we are to believe the thousands of dinosaur eggs and shell fragments preserved in the thick layers of red clay. In the temperate to subtropical climate that prevailed at the time,[8] one can imagine forests of cycas, lotus and palm trees lining the waters, with a thick underbrush of fern and lycopod plants. More temperate zones boasted pine, walnut and linden-trees.

The dinosaurs that inhabited these parts were predominantly *Rhabdodons* (bipedal ornithopods) and *Ampelosaurus* (four-legged sauropods). The *Rhabdodons* were three to four meters long, and stood on their back legs to browse on low ferns and bushes (see figure 1.3). Much taller were the great *Ampelosaurus* that could stretch their long necks into the tree canopy.

Other dinosaur species were present in lesser numbers, such as the stubby, armor-plated ankylosaurs that roamed through the bushes (see figure 1.4), and the tyrannosaur-like carnivorous giants, going by the local name of *Tarascosaurus*. Another, smaller carnivorous species left its teeth and bone fragments in the Sainte-Victoire clay. It has been named *Variraptor* by French paleontologists Eric Buffetaut and Jean Le Lœuff. These were certainly not the only dinosaur species milling around Provence at the close of the Cretaceous,

[8] At the end of the Cretaceous, the climate in Provence appears somewhat variable, as indicated by the succession of contrasting soils piled up in the sedimentary record. Some paleosoils are indicative of arid periods while others, especially at the very end of the Cretaceous, point to warmer, more humid conditions.

Figure 1.3. Dinosaur of Provence: the herbivorous *Rhabdodon* inhabited the balmy European archipelago in the Late Cretaceous. An estimated 1000 different species of dinosaurs roamed the Earth on the eve of the great mass extinction. (*Photograph: by the author, Museum of Natural History of Aix-en-Provence.*)

and rarer species, which left fewer fossils, still await discovery.[9]

Besides the dinosaurs, there were many other species of animals that shared the ecosystem in Provence's Late Cretaceous, including great lizards, turtles and crocodiles, as well as smaller amphibians and primitive mammals that fought for scraps in the underbrush. The air overhead buzzed with insects and great flying reptiles shared the skies with primitive birds.

It was this rich and diverse ecosystem that was massacred at the end of the Cretaceous period. In Provence, as on other sites on

[9] Although most of these species flourished until the very end of the Cretaceous in other islands of the European archipelago (namely in Romania), recent digs suggest that Provence and the Franco-Iberian island saw a change in dinosaur fauna before the mass extinction, with the replacement of the vegetarian sauropods and *Rhabdodons* by duck-billed hadrosaurs (see Eric Buffetaut, *Dinosaures: à la recherche d'un monde perdu*, L'Archipel, Paris, 1997).

Figure 1.4. Ankylosaurs were another family of dinosaurs that lived in Europe, and most other continents, at the close of the Cretaceous. Up to four meters long, the herbivorous ankylosaurs were studded with bony plates to deter attacks by carnivorous predators. (*Photograph: by the author, Museum of Natural History of Aix-en-Provence.*)

Earth, geologists and paleontologists long looked for clues in the terminal strata of the period that might explain the mass extinction: trends in the fossil distribution, erosional features, geochemical anomalies and any other sign of environmental disturbance. In Provence, the theory that rose to prominence in the sixties and seventies had to do with eggs.

A question of eggs

In the foothills of the Montagne Sainte-Victoire, the red clays that date back to the final days of the Cretaceous contain abundant dinosaur eggs and shell fragments, thought to belong to the great sauropods, *Rhabdodons* or hadrosaurs that came to nest in the swamps. As floods repeatedly buried the eggs in silt, layers of fossils were preserved for posterity, the older ones at the bottom and the younger ones on top.

It was this rich fossil record that French paleontologists Raymond Dughi and François Sirugue undertook to examine in the late sixties. As they sifted through hundreds of shell fragments, they discovered signs of malformation in a number of them – multi-layered shells that could have inhibited gas exchange between the egg and the atmosphere and choked the embryos. Such defects are observed in bird eggs today, when females abort laying their eggs (usually in response to stress) and recirculate them through their uterine tract where they take on extra coatings of shell.

Dughi and Sirugue claimed that the proportion of malformed dinosaur eggs increased steadily upwards in the fossil beds of the Late Cretaceous. They came to the conclusion that environmental conditions had worsened over thousands of years, disabling the reproductive functions of the dinosaurs and progressively cutting down their numbers until they vanished altogether.

Clever as it was, the malformation theory did not hold up to further scrutiny. Upon verification, there was no clear trend in the number of malformed eggs, as one ascended the fossil beds. New studies showed on the contrary that the ratio of malformed eggs remained roughly constant – a 'background' rate that was typical of the species and did not increase with time. As so many times before in the extinction debate, paleontologists were sent back to their drawing boards.[10]

The plankton crash

Whatever had caused the fall of the dinosaurs could not be told from their own fossils. Because dinosaur fossils were so rare, it was unlikely any would be found exactly at the End-Cretaceous

[10] Moreover, the Sainte-Victoire eggs were probably laid hundreds of thousands of years before the dinosaurs came to an end. More recent dinosaur fossils have indeed been discovered in southern France (see chapter 3).

boundary, and testify directly to the cause of the extinction. Instead, the breakthrough would come from the study of marine fossils.

Marine sediments of Late Cretaceous age are thicker and more widespread than their continental counterparts. They show the great mass extinction of marine plant and animal life at the end of the period to be equal to, if not worse than, that on land. Plankton, coral reefs and tributary species go extinct, as well as bony fish, ammonites, and reptiles ranging from sea turtles and crocodiles to the great mosasaurs.[11]

It is the microfossils that constitute the brunt of marine sediment, be they the millimeter-sized, plankton-like *foraminifera,* or the smaller *coccoliths* – calcareous shells of algae that are best seen under the lens of a microscope.[12] They accumulate in such numbers on the sea floor that they give paleontologists a very fine reading of events: a centimeter of marine sediment represents on the order of one thousand years of deposition, which is a very small unit of time by geological standards. This fine time resolution was of prime importance when attempting to decipher an event apparently as short as the End-Cretaceous mass extinction.

The oracle of Gubbio

One fine outcrop of marine sediment had long been studied in the hills of Umbria, Italy, near the medieval city of Gubbio. The hills in the area, gouged by deep ravines, display layer upon layer of fine limestone grading from the Late Cretaceous into the Early Tertiary. The site was part of the continental shelf at the time, lying under one hundred meters of water: a rich biological activity had snowed the sea bottom with microscopic plankton tests and

[11] The K-T extinctions in the marine realm are synchronous with those on dry land, as shown by the interfingering of terrestrial and marine sediments at K-T coastal sites, where the ocean plankton microfossils and the land-based spore fossils go extinct on the same level.

[12] These microscopic species offer crucial information about the health and diversity of marine life. They lie at the base of the food chain in the oceans, and any variation in their abundance affects the higher lifeforms.

other shell fragments, that consolidated over time into limestone, interbedded with thin layers of clay.

At Gubbio, some of the best exposures of Late Cretaceous limestone are found on the outskirts of the city, where they show up in road banks and in a nearby gorge (see figure 1.5). The limestones are pinkish and fine-grained, and their exact upper limit can be identified as a thin layer of dark clay, about one centimeter thick. Up to the clay boundary, paleontologists can easily spot Late Cretaceous microfossils with the naked eye. The darker limestones above the clay, on the other hand, lack such fossils. It takes a microscope to spot the rare and stunted 'surviving' species that mark the beginning of the Tertiary.

The thin layer of dark clay, then, represented the moment of the great mass extinction, or at least was connected to it in some way.

Figure 1.5. The first hints that the End-Cretaceous mass extinction was an abrupt crisis were discovered in the sediments of Central Italy, near the medieval city of Gubbio. There, in the limestone layers cut by a stream, a thin layer of clay separates the last fossil-rich strata of the Late Cretaceous from the fossil-poor strata of the Early Tertiary. Italian scholar Alessandro Montanari points to the emplacement of the clay boundary that became world famous as the 'K-T' layer. (*Photograph: by Walter Alvarez.*)

Because it marked this crucial boundary between the Cretaceous (symbolized in geology by the letter 'K') and the Tertiary (symbol 'T'), it became known as the 'K-T layer' to geologists around the world. Not only was this peculiar clay prominent at Gubbio, but a similar type of clay was later discovered at a number of far flung localities – from Denmark to New Zealand – as if it represented a worldwide event. It became obvious to a number of geologists that this clay was the key to the great extinction mystery.

Competing theories

At the close of the seventies, however, the significance of this K-T clay was not unanimously recognized. There were paleontologists who were still convinced that extinctions had occurred gradually at the end of the Cretaceous, and not at any specific boundary. In North America, for instance, where the K-T layer could be traced from Montana to New Mexico, a number of specialists claimed that dinosaur fossils vanished several meters below the clay layer, proof enough in their eyes that the corresponding species had gone extinct hundreds of thousands of years before the K-T boundary.

Advocates of a major catastrophe occurring exactly at the K-T boundary argued that fossilized bones of large animals like dinosaurs are rare, and that the average vertical spacing between two such bones at any dig site is typically on the order of a couple of meters. Finding a 'last' dinosaur fossil two meters below the K-T level was therefore not a proof of extinction two meters below the boundary, but a sampling bias due to rarity. Catastrophists in fact predicted that the more time paleontologists spent milling around the K-T layer, the closer to the boundary they would find dinosaur fossils.[13]

Divided as to the nature and significance of the K-T layer, paleontologists also remained divided as to the possible causes of the great mass extinction. For 'gradualists', the apparently staggered extinctions reflected a slow change in the environment, perhaps

[13] This is indeed what is observed today: dinosaur tracks and fossils are found ever closer to the K-T boundary. In 1993, geologist Chuck Pillmore reported in Colorado hadrosaur tracks a mere 37 cm below the boundary clay layer.

due to the progressive withdrawal of shallow seas off the continental shelf, or else to currents of cold arctic water flowing toward the tropics, as major shifts occurred in oceanic circulation.

For those who favored instead an abrupt, intense mass extinction at the K-T boundary, there was also a wealth of competing scenarios. Some of them called for direct cosmic interference in Earthly affairs. Indeed, ongoing studies of the Sun's nuclear cycle hinted at possible episodes of instability, when great flares of radiation would sweep through the Solar System, overrun the Earth's magnetic shield, and bombard the surface with deadly particles.

More radical yet was the suggestion that a nearby star might have exploded in the vicinity of the Earth, showering it with even stronger radiation. The probability of such a *supernova* explosion occurring within lethal range of the Earth – less than fifty light years – was by no means trivial. Such an 'outlandish' theory also had the merit of lending itself to direct verification: if the K-T mass extinc-

Figure 1.6. Few paleontologists in the 1970s were willing to consider that the dinosaurs had died out abruptly at the end of the Cretaceous. One exception was Canadian paleontologist Dale Russell, seen here during a fossil dig in the Gobi desert. Russell surmised that the cause of death was extraterrestrial and that the K-T clay might bear the evidence. (*Photograph: courtesy of the Canadian Museum of Nature.*)

tion were caused by a supernova, one would expect to find telling signs in the sedimentary record, in the form of radioactive atoms from the exploded star.

By the late seventies, geologists and paleontologists were thus faced with a broad range of scenarios to explain the great exinction of the End-Cretaceous. More pertinent data was needed to sift through the competing theories, and for many scientists, the K-T layer emerged as the oracle that would settle the issue: it would provide the answers, as long as the right questions were asked.

2

The impact hypothesis

The inquiry into the nature of the K-T boundary took a new turn in the late seventies, when a team of scientists from the University of California searched the clay for different types of rare atoms, to find out over how many years the layer had formed.

The concept of the experiment was ingenious. Rare atoms like iridium were known to be extremely rare at the surface of the Earth (less than a milligram per ton of rock) but much more abundant in the meteoritic dust that constantly rains in from outer space.[1] The concentration of 'cosmic' iridium in a sediment might then be used to indicate how long the sediment took to form – the sprinkle of stardust acting in essence as the sand of a cosmic hourglass. Measuring the concentration of iridium in the K-T clay would indicate if the clay took only a few years to form, or thousands of years of geological time.

The Berkeley team that tackled the problem was composed of geologist Walter Alvarez, his father Luis Alvarez – winner of the Nobel Prize in physics in 1968 for his ground breaking work in atomic physics – and Frank Asaro and Helen Michel at the Lawrence Berkeley Laboratory.

The team conducted its atomic test on the well preserved K-T clays of Gubbio (Italy) and Stevns Klint (Denmark), checking for over two dozen rare elements, including iridium. The results were startling.

At Gubbio, the limestones above and below the K-T boundary

[1] This average influx of meteoritic material – most of which is the size of dust specks – is estimated to reach 20 000 tons per year over the Earth's surface, i.e. a few micrograms per square meter per year. The dust is collected by scientists in deep ocean sediment and polar ice.

Figure 2.1. At many outcrops around the globe, the K-T clay jumps out as a sharp boundary between contrasting rock layers. Here at Stevns Klint in Denmark, the dark rusty clay (pointed finger) overlies the white, fossil-rich chalk of the Cretaceous. Immediately above the clay, several centimeters of darker limestone betray the thousands of years of lifeless oceans that followed the catastrophe. Higher up, the limestone regains a light-colored, fossil-rich appearance. (*Photograph: by the author.*)

showed less than 0.3 parts per billion (0.3 ppb)[2] of iridium, but concentrations in the K-T clay jumped to 9 ppb – thirty times the background average!

At Stevns Klint (see figure 2.1) the contrast was even greater: iridium reached an average concentration of 42 ppb in the K-T clay, a 160-fold increase over what it was in the surrounding limestone. One measurement even peaked at 87 ppb – a 330-fold increase of iridium as one crossed over from the limestone to the K-T clay.

Such a large amount of iridium in the K-T clay was puzzling. If one assumed that this spike reflected a longer period than usual for the centimeter-thick clay to form – allowing for all that iridium to rain in from space, as befitted the cosmic hourglass model – then the centimeter-thick K-T clay had taken more than a million years

[2] The symbol 'ppb', or parts per billion, is used for very small concentrations: it is equivalent to a tenth of a millionth of a percent.

to form! That was an absurd duration for such a thin layer of sediment. Something in the iridium 'cosmic hourglass' method was at fault, as if some extraordinary process had enriched this particular clay in iridium.

The impact hypothesis

However perplexing, the chemical analyses of the K–T clay for rare atoms like iridium had at least eliminated one theory: that of a supernova explosion.

Had the Earth been bombarded by heavy particles from an exploded star, one would have expected to find these particles in the clay, especially the radioactive element plutonium-244, otherwise absent from terrestrial sediment. The Berkeley team found no such plutonium in the K–T clay.

The supernova theory could also be independently ruled out by examining the isotopic make-up of the iridium in the clay. Iridium, like other atoms in the Universe, exists in different varieties that differ by the number of neutrons in their nucleus – varieties called *isotopes*. In our Solar System, for example, iridium exists as both 'light' iridium-191 and 'heavy' iridium-193 (with two extra neutrons), the latter being approximately twice as abundant as the former (37.3% and 62.7% respectively). These relative abundances reflect the original make-up of the cloud of dust that generated the Sun and planets. Now, if the iridium found in the K–T clay had come from another star – as would be the case in a supernova explosion – one would expect that iridium to have a different isotopic ratio, reflecting the proportions in the exploding star.

The isotopic analysis performed by the Berkeley group gave a clear answer: iridium isotopes in the K–T clay displayed the 37.3% – 62.7% ratio typical of our Solar System, and not some unusual ratio indicative of an outside source like a foreign supernova.

The enigma, however, remained whole: what was the source of the anomalous iridium, if not another star? The quantities involved were indeed staggering: if the concentrations measured at Gubbio and Stevns Klint were extrapolated to the entire surface of the

Earth – a reasonable assumption since other exposures of K-T clay (namely in New Zealand) were equally rich in iridium – then it was close to 500 000 tons of the rare metal that was dispersed around the globe during the K-T event.

The Earth itself seemed like an unlikely soure. As we already noted, iridium is extremely scarce in the Earth's crust – its concentration averaging less than 0.1 ppb. This is because early in the Earth's history, iridium seeped downward to the deep mantle and core (along with many other metals, like platinum, iron and nickel). Magmatic and hydrothermal processes do bring some iridium back to the surface, but in trace amounts: less than a ton of iridium is mined worldwide in the course of a year. It was therefore difficult to believe that half a million tons of the rare metal had been coughed up in a single event.[3] There was also no known process on Earth that could have disseminated the clay and its iridium all around the globe. A volcanic eruption, for example, was ruled out because the entire clay deposit amounted to a thousand billion tons of material, well out of range of the most cataclysmic eruptions imaginable.

Having ruled out both a supernova star and a volcanic source, the Berkeley team was left with only one solution: the Earth must have collided at K-T time with a comet or asteroid, which showered the Earth with rare metals.

It was a bold idea, and it came with a number of precisions and predictions. For instance, its authors could guess the size of the would-be 'impactor': knowing the global iridium content of the K-T layer (500 000 tons) and the average iridium content of meteorites, the Berkeley group calculated that the impacting body would have had a mass of several hundred billion tons – an asteroid or a comet about 10 kilometers in diameter.

As soon it was published in the American journal *Science* (June 6, 1980), the impact hypothesis sent a wave of consternation rippling through the scientific community. The theory was all the more impossible to ignore in that it was backed up by articles by Jan Smit, Frank Kyte and R. Ganapathy.

Dutch paleontologist Jan Smit, who had been studying the K-T

[3] Another piece of evidence rules against a terrestrial origin for the K-T iridium: its spike is not accompanied by an equal rise in nickel and chromium content, as one would expect in the case of a terrestrial, magmatic 'distillation'.

layer on the site of Caravaca, Spain, found a high iridium content in it as well. Frank Kyte reported a similar iridium spike at the K-T boundary in a deep sea core. As for R. Ganapathy, he confirmed the high iridium at Stevns Klint and also measured in the clay the concentrations of all other rare metals in the gold and platinum families. Concentrations of osmium and palladium, notably, were 1000 times greater than their terrestrial average. Moreover, when plotted together, the different metals showed relative proportions that were typical of meteorites.

Working from the assumption that the K-T clay represented a mixture of about 7% meteoritic matter and 93% pulverized target rock, Ganapathy derived a mass of 2500 billion tons for the impactor (a body 10 to 12 km in diameter), closely matching the estimate by the Berkeley team.

The precursors

Although it caught everyone by surprise, the concept of an impact devastating the biosphere was not a new idea. Throughout history, comets had often been mentioned as possible threats to life on Earth. In his book *Rogue Asteroids and Doomsday Comets,*[4] astronomer Duncan Steel reminds us that Sir Isaac Newton pondered over comet impacts as early as 1694, wondering if vast depressions like the Caspian Sea were not formed by impacts, and if the biblical Deluge was not the result of such a cataclysm.[5]

In the 1750s, French scientist and philosopher Pierre-Louis de Maupertuis went on to speculate that cometary impacts would indeed disturb the oceans and atmosphere, to the extent that many species would die out. But the concept failed to raise much interest, so little prepared were the Earth scientists to consider causes external to their planet.

It wasn't until 1956 that an American paleontologist from Oregon

[4] Duncan Steel, *Rogue Asteroids and Doomsday Comets*, John Wiley & Sons, Inc., 308pp., 1995.

[5] Newton was forced to withdraw his bold suggestion under pressure from the Church: an orbital mishap was too mechanistic a process to stand in for divine intervention.

State College, M.W. De Laubenfels, revived the issue by tackling the End-Cretaceous extinction from the standpoint of eliminated versus surviving species, in an attempt to narrow down the causes of death. In his article, De Laubenfels cited the mysterious explosion that devastated the Siberian forests of Tunguska in 1908 as an example of a collision with a cometary fragment,[6] and compared the event with the trauma of the End-Cretaceous.

According to the scientist, a major impact would cause brief but extremely brutal rises in temperature. Fires would ensue, but much of the vegetation would survive in seed form, which would explain why the fossil record showed only minor extinctions in the plant kingdom. Animals, on the other hand, would have been much more vulnerable to heat. De Laubenfels speculated that those terrestrial species which did survive were small, burrowing animals that waited out the heat wave in their 'insulated' burrows. As for the fresh water fauna, the survival of most amphibians, like frogs and turtles, could likewise be explained by their ability to dive underwater and stay near the cool bottom for the duration of the heat blast.

De Laubenfels' speculations were challenging, but again they were ignored by paleontologists. It was another fifteen years before more light was shed on the matter, in a 1972 article written by Harold Urey. The Nobel prize-winner made the bold suggestion that not only the End-Cretaceous but most other mass extinctions on record were due to cometary impacts. Urey estimated that a large comet impact would trigger a surge of temperature in the atmosphere on the order of 200 K. The physicist also pointed out that such impacts could be confirmed by the discovery of tektites.

Tektites are vitreous rocks with rounded shapes, including tear and dumb-bell profiles. In the seventies, the emerging consensus was that they were drops of impact melt sprayed about by impacts, that spun and solidified into their rounded shapes during their brief flight through the atmosphere.

Harold Urey realized that such glassy fragments would constitute primary evidence for an impact at an extinction level, such as the

[6] For a discussion of the Tunguska impact, see the first section in chapter 8.

K-T boundary. But at the time of his article, no such tektites had been signaled at extinction boundaries. It would be another eight years before Alvarez *et al.* presented the first evidence of an impact at the K-T boundary, in their 1980 landmark paper.

The wind turns

Following the 1980 articles by Alvarez, Ganapathy, Smit and Hsü, research blossomed on the K-T extinction and the cosmic hypothesis. The thin boundary clay was identified and studied in a growing number of locations. After Gubbio (Italy) and Stevns Klint (Denmark), new sites were explored in Spain, Tunisia, New Zealand and the Americas. Independent iridium measurements were conducted by European geochemists, and notably by Robert Rocchia of France's CNRS/CEA laboratory at Gif-sur-Yvette, confirming earlier results (see figure 2.2). By 1982, close to forty sites with iridium anomalies were reported worldwide, spanning Oceania, Africa, Europe and Asia, and the Americas from Montana to the Caribbean basin.

To their disappointment, geologists did not find any trend in the iridium concentration from site to site that would point to the location of the source crater. The Caravaca site in Spain, the site in New Zealand, and a submarine site near Hawaii were all equally rich in iridium. Apparently, the emplacement of the iridium-rich clay was a global mechanism that left no clue as to the location of ground zero.

Meanwhile, new clues were found in the K-T layer that confirmed the theory of a cosmic impact. Dutch paleontologist Jan Smit found that in southern Spain the clay contained abundant small 'droplets' about a millimeter in size that strongly resembled – albeit on a diminutive scale – the tektites that Harold Urey had recommended searching for at extinction boundaries. Could these 'microkrystites', as Smit called them, indeed be minuscule droplets of impact melt?

Figure 2.2. French geochemist Robert Rocchia, retrieving samples of K-T clay in southern Spain. The centimeter-thick clay that marks the moment of the mass extinction is lodged at the top of the sample, level with Rocchia's right thumb (arrow). (*Photograph: by Eric Robin, CEA/CNRS.*)

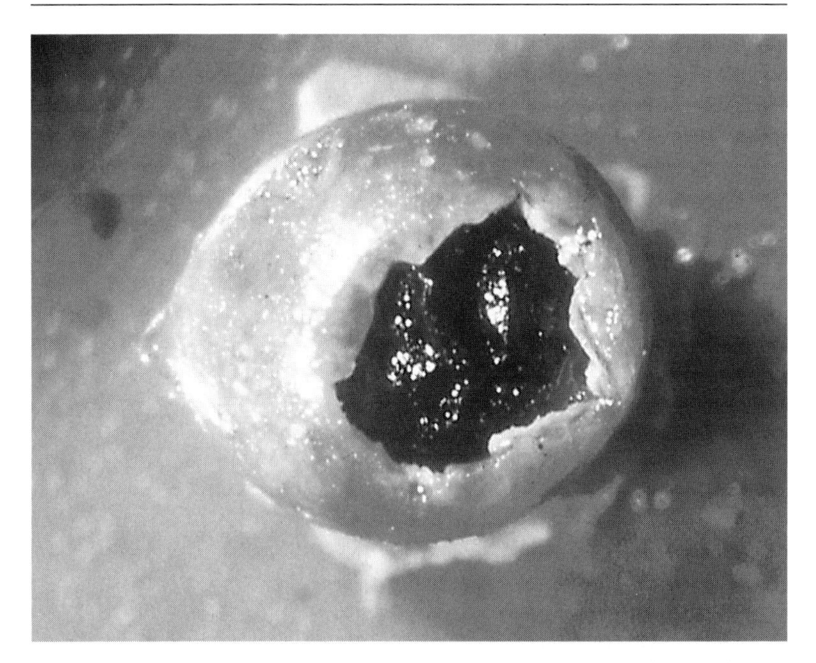

Figure 2.3. Spherules of impact glass are crucial evidence of a cosmic collision at the K-T boundary. Known as tektites, these droplets of ejecta are sprayed into the atmosphere during impact and cool as they stream back to earth, taking on aerodynamic shapes like spheres and teardrops. This sample from Haiti is two millimeters across and shows a core of dark impact glass, coated by a layer of clay. (*Photograph: by Glen A. Izett, U.S. Geological Survey, Emeritus.*)

Shocked quartz

The iridium anomaly and the presence of other rare metals in meteoritic proportions were already good evidence of a cosmic impact at the K-T boundary, at least to geochemists (see figure 2.4). But, as we shall see in the next chapter, that evidence was long challenged by the majority of geologists, who were unfamiliar with geochemistry. Also challenged were the microkrystites reported by Jan Smit, because their resemblance to tektites was far from conclusive, and their chemical make-up was too altered to

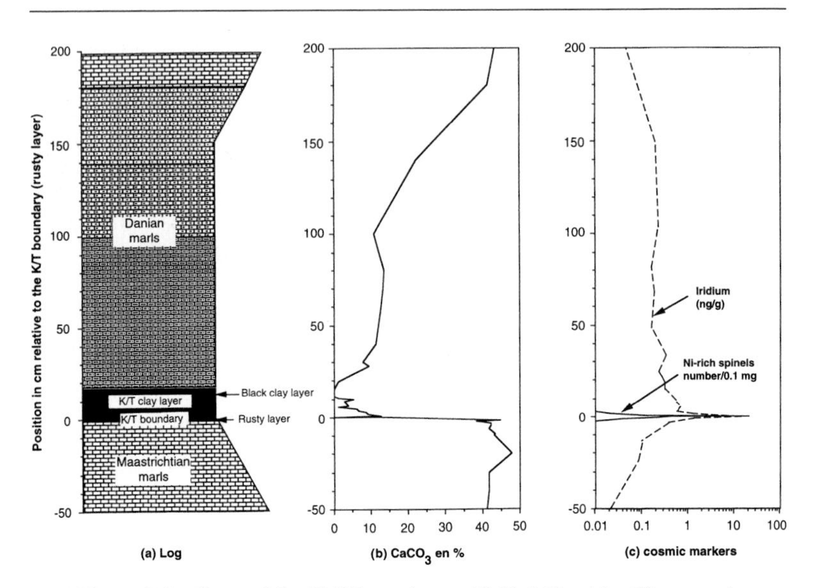

Figure 2.4. Chart of the K-T boundary at El Kef, Tunisia. The organic content of the sediment, as represented by its $CaCO_3$ content, drops dramatically at the clay boundary, and returns to normal values a full two meters above the boundary – i.e. after hundreds of thousands of years of slow recovery of the biosphere (middle column). In the column at right are plotted the clues of a cosmic impact: the iridium concentration (dashed line) peaks at the level of the K-T boundary, as does the number of crystals of meteoritic spinel which are entirely contained in the centimeter-thick clay. (*Courtesy of Robert Rocchia and Eric Robin, CEA/CNRS.*)

yield clear evidence of their origin. In fact, during its first four years of existence, the impact theory was sorely in need of a legitimate confirmation – a classic proof of impact unanimously recognized by specialists in the field.

One such proof would be minerals shocked at very high pressures. Shocked minerals were known to be symptomatic of impact. Collisions generate such high pressures at ground zero that the rocks which escape melting are brutally shattered, and their minerals undergo dislocation and vitrification in the form of criss-crossing bands ('lamellae') that can easily be spotted under the microscope. Quartz grains showing such multi-lamellar deformation are only found at meteoritic impact sites, and in the cavity walls of underground nuclear test grounds.

To validate the impact hypothesis, specialists were therefore anxious to find shocked minerals at the K-T boundary. For reasons discussed in the next chapter, the search was long delayed, but finally met with success in 1984: geologist Bruce Bohor and his colleagues of the U.S. Geological Survey reported the discovery of shocked quartz on the K-T site of Hells Creek in Montana.

On this continental site, east of the Rockies, the K-T clay was already known to contain abundant iridium (1 ppb, about 200 times the local norm), as well as microkrystites of feldspar, reminiscent of tektite forms. But it was the quartz that caught the attention of Bruce Bohor and co-workers.

Figure 2.5. Geologist Bruce Bohor of the U.S. Geological Survey (at right) was first to report shocked quartz in the K-T clay, a clue which greatly boosted the impact hypothesis. He is seen here on the K-T site of El Peñon, Mexico, along with French geochemist Eric Robin (left). (*Photograph: by the author.*)

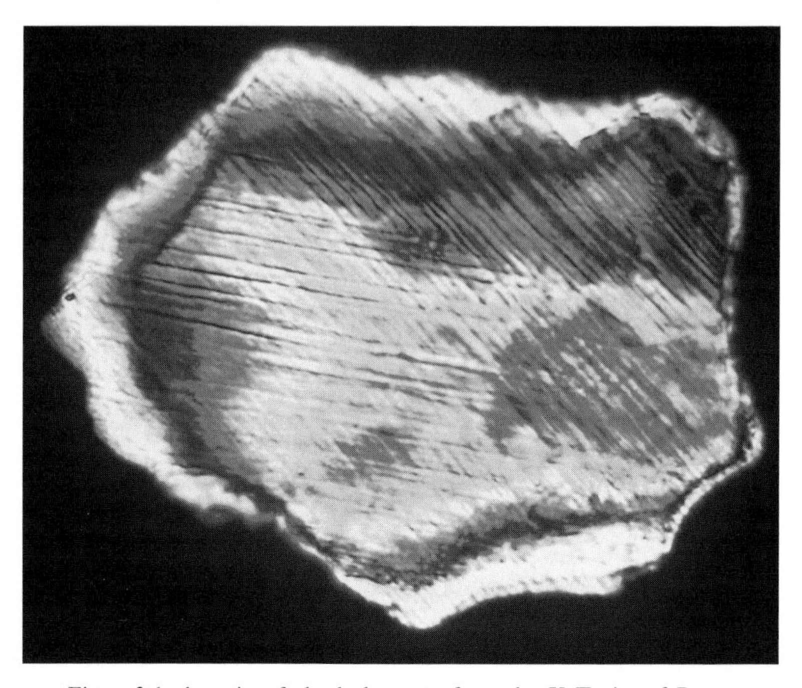

Figure 2.6. A grain of shocked quartz from the K-T site of Raton Basin, Colorado, seen under the polarizing microscope. A quarter of a millimeter in size, the mineral grain displays two sets of lamellar deformations, criss-crossing at an oblique angle. This hatching pattern is typical of shock at very high pressures, and can only be produced by nuclear explosions and meteoritic impacts. (*Photograph: by Glen A. Izett, U.S. Geological Survey, Emeritus.*)

Examined under the microscope, nearly a quarter of these quartz grains were criss-crossed by the inimitable lamellar features produced by strong shockwaves (see figure 2.6). By comparing the deformation pattern in the K-T quartz grains with those obtained in laboratory experiments, the scientists deduced that the K-T quartz was exposed to pressures on the order of 10 gigapascals, corresponding to very large impacts.[7]

[7] At the high pressures of 10 gigapascals and more, quartz can change structure entirely, compacting into new silica crystals named coesite and stishovite. Stishovite is only known at impact craters and on nuclear sites. It was discovered at the K-T boundary in New Mexico by J.F. McHone and R.L. Nieman in 1989.

The landmark report by Bohor *et al.* sparked a flurry of new discoveries. In three years' time, no less than nine other K-T sites were found to contain shocked quartz. Bruce Bohor and his coauthors reported seven of these, including the landmark sites of Gubbio and Stevns Klint. Glen Izett and Chuck Pillmore, also of the U.S. Geological Survey, signaled a site with iridium and shocked quartz in New Mexico, and Soviet geologist Badjukov reported a site with iridium and shocked quartz in Russia.

In all studied samples, the number of shocked quartz grains[8] and their degree of deformation were remarkable. Over a quarter of the quartz grains displayed multiple planes of dislocation, as well as other signs of high pressure.

The shocked quartz therefore brought a strong, independent confirmation of an impact at the K-T boundary. But, just as importantly, it offered an important clue as to the location of the crater itself.

Initially, most adepts of the impact hypothesis favored an oceanic impact. It was the most probable choice, since oceans cover two thirds of the Earth's surface. But the shocked quartz told a different story. Quartz of such size is typical of continents, and is extremely rare in the ocean crust. The occurrence of quartz in the K-T clay thus hinted that the impact had taken place on a continent. Whereas scientists had almost given up hope of finding a 65 million year old crater on the sea floor, the prospect of finding a crater on the continent looked much brighter.

A sprinkle of diamonds

Besides iridium and shocked quartz, more evidence of impact was to be found in the K-T layer. In August of 1991, Canadian scientists Carlisle and Braman reported microscopic diamonds in the K-T clay of Alberta.

Probing the clay with an electron microscope, the scientists

[8] Further field work and analyses reveal that quartz is not the only mineral to be shocked in the boundary clay, although by far the most abundant. Feldspar, chromite and zircon also show crystal deformations characteristic of high velocity impacts.

found small mineral grains three to five nanometers in size – smaller than viruses – that matched the known spectroscopic signature of diamond.

The discovery was confirmed the following year by a British team led by Iain Gilmour, that reported microdiamonds in two other K-T sites: in Montana and Colorado. These were also close to five nanometers in size.

Such minuscule diamonds had no commercial value, but they told a precious story. Diamonds are carbon crystals that form at very high pressures. Most diamonds are thought to originate tens of kilometers underground, where carbon is slowly compressed by the weight of the overlying rock. These deep rocks are occasionally brought up to the surface through explosive volcanic eruptions.

The diamonds of the K-T layer, however, do not speak of underground formation. They are minuscule, and such small crystals would have turned to carbon vapor if they had been exposed to the high temperatures of volcanic eruptions. Moreover, the K-T microdiamonds are almost free of nitrogen impurities, contrary to the 'dirtier' diamonds originating in the Earth's mantle.

The Earth's interior was ruled out, but there were two other ways of accounting for the diamonds. Either they had come from space – small diamond crystals are found in chondritic meteorites – or else they were formed at the Earth's surface by the high pressure and temperature of an impact, as was notably the case at Popigai crater in Russia.

Several lines of evidence support the latter interpretation. Iain Gilmour and co-workers point out that however minuscule, K-T diamonds are still larger than those found in meteorites (about double the size), and moreover they display carbon isotopic ratios (C-13 versus C-14) that are significantly different from those of meteoritic diamonds. The scientists conclude that the K-T diamonds did not exist prior to impact, and that they were formed from carbon supplied by the meteorite or by the target rock.[9]

Carbon is also found in the K-T clay in the form of amino-acids. Amino-acids are intricate chemical compounds of carbon, nitrogen,

[9] Diamond formation apparently took place either by direct shock alteration of solid material on the impact site, or by condensation of diamond crystals out of the chemical vapor lofted by the fireball.

Figure 2.7. Eric Robin, of France's CEA/CNRS lab, at work on the K-T site of Mimbral, Mexico. He points to the layer in the sediment where the cosmic iridium is concentrated. Microscopic diamonds can also be found in the clay, created by the high pressures and temperatures of the impact. (*Photograph: by the author.*)

oxygen and hydrogen. There are many varieties of amino-acids in the Universe: some are typical of chemical processes on our planet (and among these, a few are the building blocks of life as we know it); and others are found only in 'raw' cosmic material, namely in meteorites. It so happens that those of the K-T layer belong to the latter. At Stevns Klint in particular, one finds isovalin and alpha-amino-isobutyric acid, cosmic varieties which are not normally found on Earth. On a site in Canada, Carlisle and Braman found 51 different types of amino-acids, 18 of which have no equivalent on Earth. In fact, on the Stevns Klint site, these fragile molecules are found not in the clay itself but immediately above and below the boundary layer, which led Kevin Zahnle and David Grinspoon of NASA to suggest that the incoming projectile from space was a comet rather than an asteroid, raining its cosmic matter on Earth well before impact, and leaving a trail of dust in orbit that continued to fall long after the core had hit the Earth.

The testimony of spinels

There is one last line of evidence that signs away an impact at the K-T boundary: the abundance of spinel crystals.

Spinel is the name of a vast family of metal oxides – opaque minerals that enclose variable amounts of iron, magnesium, aluminum, titanium, nickel and chromium. Those found in the K-T clay are both plentiful and distinct. Under the microscope, they display elegant shapes, including snowflake, octahedron and cruciform motifs (see figure 2.8).

These dendritic shapes are characteristic of rapid solidification out of a high-temperature melt, and their chemical composition is unusual. Terrestrial spinels are created underground in volcanic magmas, where they are cut off from atmospheric oxygen. In this low oxidation state, they are characterized by high proportions of iron, titanium and chromium.

Spinels of the K-T layer, on the other hand, are rich in nickel and magnesium, which is the sign of a high oxidation state. Such highly oxidized spinels are known to form only through *ablation*, the melting of a metal-rich body diving through the Earth's oxygen-rich atmosphere.

The process is familiar to meteorite experts: during their fall to Earth, cosmic objects are heated on the outside, and droplets of fused material shed off the surface and recrystallize into new minerals in the oxygen-rich atmosphere.

Because a good deal of meteoritic matter is nickel-rich (chondritic meteorites contain 1 to 2% nickel), the ablation process typically generates vast amounts of nickel-rich spinel. Since the process takes place in an oxygen-rich atmosphere – made even denser by the compression factor of the incoming projectile – these spinels attain very high oxidation states.

Geochemists can even guess, from the degree of oxidation, at what altitude the ablation took place. Oxidation states around 60%

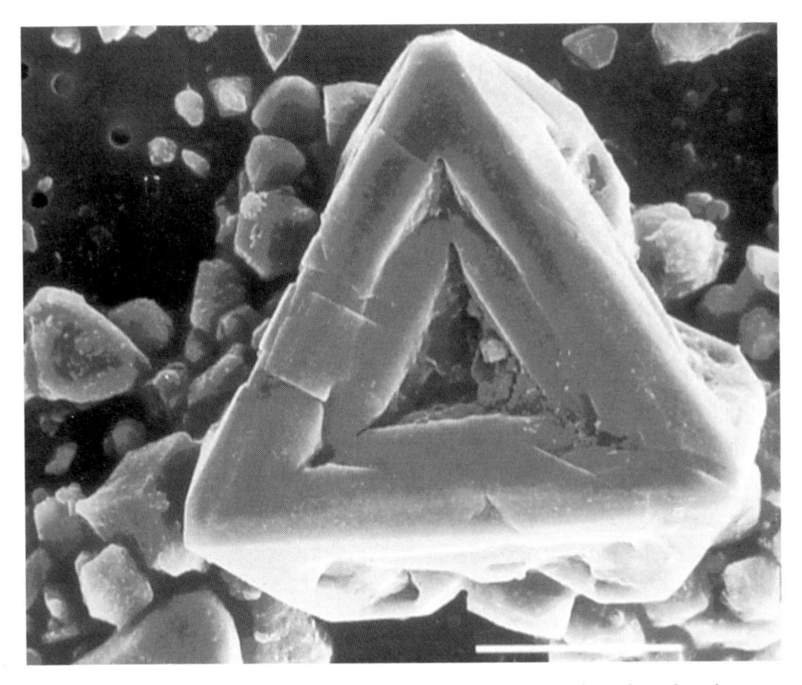

Figure 2.8. A crystal of spinel from the K-T clay, viewed under the electron microscope. These nickel-rich minerals can only be formed by the melting and oxidation of meteorites as they enter the Earth's atmosphere, and offer unambiguous proof of an impact at the K-T boundary. (*Photograph: by J. Gayraud and E. Robin, CEA/CNRS.*)

indicate that the meteorite was decelerated high up in the atmosphere, between 40 and 80 km, where there is relatively low oxygen pressure. To be decelerated so fast means that the projectile did not have much momentum, and thus had low mass. Spinels of higher oxidation point to ablation much closer to the ground, where oxygen pressure is highest. Only large, massive 'bolides' penetrate deep enough into the atmosphere to get their vapor droplets oxidized to such an extent.

The spinels of the K-T layer indicate from their oxidation state (80 to 100%) that their altitude of formation was on the low side, below 20 km. Thus the ablation crystals were shed off a large projectile – or projectiles – penetrating deep into the oxygen-rich atmosphere.

The K-T spinels also bring another piece of information to the puzzle. They confirm that the deposition event that created the clay was very brief.

The original iridium data had left some scientists skeptical about the brevity of the K-T event, because iridium was weakly spread out on both sides of the K-T clay layer, and could give the idea of an event spread out over tens of thousands of years.[10] The spinel crystals told otherwise. They were concentrated only in the few millimeters of the K-T clay, near the base. On the K-T site of El Kef in Tunisia, geochemist Eric Robin and his colleagues from the French National Research Center noted that 95 % of the spinel lay in the lower two millimeters of the clay – the mark of a near instantaneous cosmic event (see figure 2.4).

Coming after the evidence of iridium, shocked quartz, microkrystites, diamonds and amino-acids, ablation spinels put the icing on the impact hypothesis. All this, as we shall see in the next chapter, came under the constant scrutiny and challenge of dissenting opinion.

[10] In fact, it now appears that the iridium 'smear' on either side of the K-T clay is due to chemical alteration of the sediment, causing the atoms in the K-T clay to diffuse outward, and biological activity – 'gardening' and mixing of the soil by burrowing animals like worms.

3

The controversy

When it was first published in 1980, the cosmic hypothesis met with a solid barrage of criticism. The opposition stemmed mostly from the fact that the event was portrayed as having been instantaneous, and extraterrestrial.

Instantaneous and violent events were frowned upon in the Earth sciences. In ‘mainstream’ geology, as we saw in chapter 1, the prevalent model was that all processes were slow and gradual. As for the extraterrestrial aspect of the question, it was nonsense to scientists who liked to reason within a terrestrial, closed frame of reference. There was no need to call upon a cosmic intervention in order to settle a ‘domestic’ issue on Earth. Most geologists had little experience of cosmic processes in the first place. Impacts were not part of the curriculum, except in rare courses of planetary geology that were offered by very few colleges and universities.

There was another reason for the hostility towards the cosmic hypothesis: its sheer simplicity was irritating. Paleontologists and geologists had worked in the field and argued for decades to try to resolve the mystery of mass extinctions, and along came a crew of ‘laboratory’ physicists who pretended to put the whole issue to rest with one wave of their cosmic wand. In this context, it was not a surprise that the thesis of an instantaneous cosmic catastrophe was met with overt skepticism.

The volcanic hypothesis

At first, the cosmic nature of the iridium was challenged. Since the early samples of K-T clay were formed in marine environments, geologists rightly questioned whether, instead of being cosmic, the iridium wasn't simply marine in origin, and if its diluted content in sea water had not been concentrated and dropped to the bottom by some unknown mechanism.

The marine alternative did not hold up to scrutiny. For one, even if all the iridium in a three kilometer column of water was precipitated to the ocean bottom, one would still be one hundred fold short of the iridium concentration found in the K-T clay.[1] In addition, iridium-enriched clay was soon discovered in continental strata as well, in sediment that had formed in freshwater environments, away from the sea. The iridium anomaly was therefore not a marine artefact, but truly a planet-wide phenomenon.

Next came a formal challenge of the cosmic hypothesis by Charles Officer and Charles Drake, geology professors at Dartmouth College, who put forth a volcanic scenario to explain the presence and dispersal of iridium around the globe.

The idea of a volcanic catastrophe rocking the Late Cretaceous was not new. Peter Vogt had suggested climate-disturbing eruptions as a cause of mass extinctions as early as 1972, and the concept was further explored by Dewey McLean in a paper published in 1975.

In their new scenario, Officer and Drake postulated that the K-T iridium was volcanic in nature. Since iridium is found in much larger quantities in the Earth's mantle and core than in the crust, the anomalous iridium could simply have been pumped up to the surface by deep-rooted volcanoes (of the 'hot spot' variety) and blown high into the atmosphere, where strong winds would have carried it around the globe.

Drake and Officer cited Hawaiian volcanoes as an example. Measurements taken by chemist William Zoller and his team at Kilauea crater indeed showed unusual concentrations of iridium in the airborne particulate matter collected downwind from the vent (up to three grams of iridium per million cubic meters of volcanic fumes).

It thus appeared that a fair percentage of iridium could leave the magma during an eruption and become concentrated in the

[1] Sea water contains around 10^{-15} g/g of iridium. If one purged a three kilometer column of water of all its iridium, one would obtain 0.3 nanograms per cm^2, less than one hundredth the quantity observed in the K-T layer.

Figure 3.1. A fountain of lava at Kilauea volcano, Hawaii. As a counterpoint to the impact theory, it was suggested in the mid-eighties that an episode of deep-rooted volcanism could also account for the iridium found at the K-T boundary. (*Photograph: by J.D. Griggs, U.S. Geological Survey.*)

airborne particles. However, this did not change the fact that the global concentration of iridium in Hawaiian magma was low to start with – approximately 0.5 ppb at Kilauea – so that even if this iridium became noticeable in the airborne particulate fraction, it would still take *millions* of eruptions like the ongoing one at Kilauea to belch out the 500 000 tons of iridium found in the centimeter-thick K-T clay.

Besides the issue of quantity, the volcanic explanation for the iridium was further weakened by geochemical considerations. In the K-T clay, iridium was accompanied by other rare metals – such as gold and rhenium – in proportions that were typical of meteorites, not volcanic lavas. Hence, with respect to iridium, the volcanic scenario was much less convincing than the cosmic one.

Skeptics, and notably 'volcanists'[2] as they came to be known, next reviewed the evidence of shocked quartz. These deformed crystals are characteristic of pressures that can only be produced in meteoritic impacts and nuclear explosions, as we saw in chapter 2. But this diagnostic characteristic was challenged by Officer, Drake and others, who claimed that volcanic explosions could also deform crystals to the same extent, citing as examples the 'volcanic intrusions' of Vredefort in South Africa, and the Sudbury complex in Canada, where abundant shocked quartz can be found.

But Vredefort and Sudbury were the wrong pick: they were found to be impact craters! In later articles, Charles Officer and coauthors accordingly retracted their reference to Sudbury, and minimized their reference to Vredefort, to center their claim on the shocked minerals of a true volcano – Toba in the island of Sumatra, a gigantic caldera 100 km long by 30 km wide, which held its last cataclysmic eruption 75 000 years ago.

To their credit, the volcanists did find some evidence of shock deformation at Toba caldera, but in one quartz grain out of a hundred, and with dislocations running only along a single plane, which did not portend to very high pressures. This was in strong contrast with the shocked quartz grains of the K-T layer, which were outstanding both in terms of quantity – 25% of the K-T quartz grains in North America were shocked – and intensity of deformation, characterized by multiple criss-crossing planes. But despite the strong differences, Officer *et al.* stuck to their claim that the K-T quartz was volcanic.

The Deccan Traps

Although their arguments were frail, 'volcanists' had an appealing candidate that they could claim responsible for the K-T mass extinction: the Deccan Traps of India.

[2] It is interesting to note that the 'volcanists' (supporters of a volcanic origin for the K-T layer) were not volcanologists (specialists in volcanoes). The latter did not endorse their views.

A *trap*[3] is an impressive accumulation of basalt, consisting of many lava layers piled up to form extensive plateaus. They are also called continental flood basalts – 'flood' because their eruptive rates are extremely high and spread the lava over large areas. A dozen such lava fields were emplaced on Earth over the past 300 million years, comparable in scope to the *maria* of the Moon or the vast volcanic plains of Mars and Venus.

The Deccan Traps of India (see figure 3.2) cover more than 500 000 km^2 – the area of a country like France – and their thickness averages 2000 meters. Thus, their total volume is estimated to be around one million cubic kilometers of lava.

The Deccan Traps are exceptional not only in terms of volume but in terms of age as well: stratigraphy, geomagnetic dating and radiochronology indicate that the main sequence of eruptions took place in the Late Cretaceous and Early Tertiary, straddling the K-T boundary. The coincidence was intriguing enough to raise the issue of a causal relationship.

The link between the Deccan Traps and the K-T mass extinction had been suggested by Peter Vogt in 1972, and was revived by Officer and Drake in the mid-1980s, as well as by geophysicist Vincent Courtillot in France who had been studying the lava flows of India in the course of a broad geomagnetic survey. The French scientist and his team were especially intrigued by the paleomagnetic readings of the lava, which hinted that the main sequence of eruptions in the Deccan Traps had taken place in less than a million years – a much shorter interval than previously thought.[4]

The model they reached was that brief and violent eruptions polluted the atmosphere with ash and brought drastic climate

[3] The word *trap* comes from the Swedish (and Danish) word *trappa*, meaning 'stairs': indeed, the lava piles are often eroded in a terraced, step-like fashion. Recent examples of traps on Earth are the Ethiopian lava fields, which preceded the opening of the Red Sea, and the Columbia River Plateau of the northwestern United States which erupted 17 to 15 million years ago.

[4] Volcanic rocks, when they cool, are known to 'freeze' an imprint of the prevailing magnetic field – in particular its polarity (north–south or south–north). The polarity of the Earth's field has switched repeatedly through time in a unique pattern that provides a relative timescale. The K-T boundary, for instance, is known to have occurred during a period of reversed magnetic polarity, known as period '29R', which lasted approximately 500 000 years. According to Courtillot *et al.*, the main eruptions in the Deccan Traps spanned only three polarity intervals – the end of 30N, the whole of 29R and the beginning of 29N – i.e. less than a million years.

Figure 3.2. The Deccan Traps of India. These huge outflows of basalt cover an area the size of France and their piles of alternating dark and light layers are over a thousand meters thick. Dated around 65 million years old, the Deccan eruptions were postulated by some to be the cause of the K-T mass extinction, by upsetting the climate at the end of the Cretaceous. (*Photograph: by Robert Rocchia.*)

changes that caused the mass extinction. In this view, the K-T clay represented a violent paroxysm in the midst of the Deccan eruptive cycle (complete with volcanically emitted iridium and explosively emplaced shocked quartz), bringing the *coup de grâce* to a declining biosphere.

This volcanic model, again, went against the standing evidence that the iridium and shocked quartz of the K-T layer were impact-produced and not volcanic. To further weaken the matter, an analysis of the Deccan lava by geochemist Robert Rocchia and his team showed it to be particularly lacking in iridium, with concentrations lower than 0.01 ppb. Even if one extracted the totality of the iridium present in the two kilometer-thick pile of basalt, one wouldn't reach one hundredth of the quantity of iridium disseminated worldwide in the K-T layer. Neither was there a single grain of shocked quartz in the Deccan Traps, which was not a surprise since quartz is notably rare in basalt.

Only the timing of the Deccan Traps and its narrow coincidence with the K-T crisis gave reason to ponder, but even that evidence was not conclusive. Courtillot's geomagnetic dating pinned the bulk of the eruptions down to a timeframe of less than a million years, centered around the K-T boundary, but other work by T.R. Venkatesan and his co-workers at India's Physical Research Laboratory, based on radiometric dating, hinted that the eruption rates in the Deccan peaked not 65, but 67 million years before present, a full two million years before the K-T crisis.[5] Moreover, the most polluting eruptions of the Deccan Traps are thought to have occurred even before that, during the opening phases of the continental rifting in India, when crustal failure and the engulfment of siliceous material in the magma might have led to explosive eruptions. And yet, it is only at K-T time that the mass extinction took place, well over two million years later.

Confirming this offset in age that contradicts a true causal relationship, Indian geologist Narendra Bandhari appears to have found the local expression of the K-T boundary not at the bottom but high up in the sequence of the Deccan Traps, in the middle of a nine meter thick layer of sediments, sandwiched between two late lava flows in the western Anjar province. Apparently then, the K-T event took place towards the end of the volcanic sequence, and during a local hiatus in the activity, to boot!

To close the issue, it is worthwhile to ask the dinosaurs themselves what they thought of the Deccan eruptions. Indian geologists Z.G. Ghevariya, S. Bajpai and their teams have found abundant dinosaur bones and egg shell fragments in the Anjar sediment between the lava flows, straight up to the K-T boundary (and none above it). Obviously then, dinosaurs were thriving and reproducing in the Deccan Traps, quite unaffected by millions of years of

[5] The Deccan eruptions probably lasted more than five million years. Indeed there are layers of volcanic rock (about 1500 m thick) hidden underwater in the Cambay Basin that predate the 67 m.y. old flows visible on the mainland. Similarly, layers younger than 62 m.y. were apparently eroded away from the top of the Mahabaleshwar Formation, south of Bombay.

volcanic activity. Because India was an island at the time, these dinosaurs could not possibly have fled during major eruptions to later return when things calmed down. The fact that they survived on the very site of the eruptions says a good deal about how 'catastrophic' these eruptions really were.

Gradualism revisited

Throughout the controversy, many opponents of the cosmic theory kept on claiming that the K-T mass extinction had not been sudden but had spanned hundreds of thousands, or millions of years. By refuting the concept of a sudden extinction, they could do away with the whole idea of an impact. In fact, the K-T layer itself could be dismissed as irrelevant – along with its nagging cosmic evidence – since there would be no more brutal extinction for it to explain.

The line of argumentation was not new and had considerable support: many paleontologists challenged the brevity of the mass extinctions, and supported scenarios of drawn out volcanic eruptions changing the climate, or of sea regressions shrinking the continental platforms and stressing the biosphere.

The fossil record of the dinosaurs, in particular, remained a major issue. In his book *The Great Dinosaur Extinction Controversy,*[6] volcanist Charles Officer holds the extreme view that 'dinosaurs disappeared outside North America well before the Cretaceous came to an end . . . And in North America, the dinosaur demise began as a gradual process . . . The two last known dinosaur species were the herbivore *Triceratops* and the carnivore *Tyrannosaurus*.'

Such claims are widely disputed. In Montana alone, most paleontologists would agree that there were between 10 and 15 different dinosaur species roaming its alluvial plains at the close of the Cretaceous. Some even put the regional number at 20, and stress that the difficulty of finding fossils of large vertebrates certainly leads to an underestimate, so that the real number of dinosaur

[6] *The Great Dinosaur Extinction Controversy*, by Charles Officer and Jake Page, Addison-Wesley Publishing Company, Inc., 1996.

species in Montana was probably double the official figure and closer to 40.

Until recently, there were no dinosaur dig sites outside North America that came close enough to the End-Cretaceous boundary to enrich the discussion, but the more paleontologists look, the more sites they discover. In southern France, at the foot of the Pyrenees Mountains, vertebrae of hadrosaurs (duck-billed dinosaurs) have been unearthed in Late Cretaceous strata, less than a meter below the K-T boundary. In India, as we saw in the previous section, bones and egg shells of sauropod dinosaurs are found in the Deccan Traps, very close to the K-T layer.

Many more finds are expected in the near future as paleontologists explore End-Cretaceous sites in China, the Sahara and South America. In fact, given the diversity of habitats around the world, paleontologist Dale Russell estimates that there were close to *one thousand* different species of dinosaurs roaming the Earth at the close of the Cretaceous – a lively bunch that prospered all the way to K-T time.

Of course, there were other animal species that did seem to decline in diversity long before the K-T boundary, and testified to at least some gradual extinctions at the close of the Cretaceous. The ammonites were one such group.

Ammonites were predatory sea creatures – large mollusks with tentacles, protected by coiled, spiral shells, drifting and feeding in the open waters of the great world oceans. They came in all shapes and sizes, some reaching up to one meter in diameter. The number of different ammonite species at any one time ranged in the dozens.

At the close of the Cretaceous, this diversity seemed to decline. As he combed over the fossil beds of northern Spain, ammonite specialist Peter Ward reported in the mid 1980s that the ammonite species progressively petered out over more than 100 meters of strata – a drawn-out interval of at least one million years. This was hardly a brutal mass extinction and supported the view of gradual climate or sea level change, killing off the species one by one. Moreover, the very last ammonite fossil was found more than ten meters below the K-T boundary, hinting that the last of the ammonites had died 100 000 years before the postulated catastrophic impact.

Figure 3.3. Geologist Charles L. Pillmore rests against the natural casting in sandstone of a T-Rex footprint, showing its three prominent toes. Pillmore found the fossilized track less than a meter below the K-T boundary in New Mexico, confirming that dinosaurs did not go extinct long before the boundary, but were thriving right up to the impact. (*Photograph: courtesy of Charles L. Pillmore, U.S. Geological Survey, Emeritus.*)

But just as was true of the dinosaur record, it was the scarcity of ammonite fossils – especially those of rare species – that made the extinction look gradual. The more scientists milled over the rocks in the vicinity of the K-T boundary, the more ammonites they found. Geochemist Robert Rocchia found two ammonite imprints 20 to 30 centimeters below the K-T layer in northern Spain.[7] In fact, when Peter Ward returned to the Spanish outcrops and covered new ground, he found ammonite upon ammonite in the last meter and a half of limestone below the boundary – a total

[7] This comes as a second paleontological find for Robert Rocchia. The geochemist also discovered dinosaur vertebrae in Late Cretaceous strata of southern France.

of 42 fossils, representing 12 out of the 28 species reported in the older strata.

Summarizing the data and using statistical methods to estimate the true level of extinction of each species (which is always somewhere above the last fossil occurrence), Peter Ward and Charles Marshall concluded that six ammonite species did go extinct long before the K-T boundary in a random fashion (background extinctions, which occur all the time), and that perhaps as many as ten species disappeared from the area 20 000 to 10 000 years before the K-T boundary, probably because of a sea level drop that shrunk the area of the continental shelf and increased the level of competition between animals, as some gradualists had rightly claimed. But at least twelve ammonite species had been alive and well until the very last days of the Cretaceous, going out with a bang at precisely the K-T boundary. Even ammonites, then, were found to have suffered a catastrophic mass extinction at the boundary.

Before a jury of peers

Interpreting the fossil record of large animals, like dinosaurs and ammonites, was controversial because they left relatively few fossils to study. The abundant fossils of small species like plankton, on the other hand, were believed to provide a much more accurate and unambiguous record of the K-T crisis.

But such was not the case. A number of micropaleontologists, like Jan Smit of the Free University of Amsterdam, saw an abrupt mass extinction of plankton species right at the K-T boundary, whereas others, like Gerta Keller of Princeton, saw a stepwise, gradual extinction of the same species over tens of thousands of years up to, and across, the boundary. Clearly, something – or someone – was wrong.

In order to settle the issue, gradualists and catastrophists took the bold step of bringing their case to trial, by setting up an impartial test to be performed by a jury of experts. In this 'blind test', samples were collected at the fossil-rich site of El Kef in Tunisia, at six different horizons below and above the K-T boundary, and were distributed to a panel of four paleontologists, who set out to identify the plankton species in each. For the test to be as impartial as possible, the jury were given the batch of samples in random order, without any indication as to which level – above or below the boundary – each sample represented.

The test was organized under the supervision of 'judge' Robert Ginsburg, sedimentologist at the University of Miami, who compiled the results. These were first presented at the Houston K-T symposium in February of 1994. Before the oral presentation of the results, gradualist Gerta Keller and co-workers claimed victory in a posted report, where they pointed out that each of the investigators had found species to be missing in samples from below the K-T boundary, proof in their eyes that these had gradually gone extinct.

But the catastrophists came up with a different interpretation. Jan Smit pointed out that each member of the jury had indeed missed one or several species of plankton in each sample – rare species that other jurors had otherwise signaled in their samples from the same level. This simply illustrated how difficult it was to identify or find a rare species. A paleontologist working alone could believe in the disappearance of a species for having overlooked it, mistaken it for another, or simply because it was lacking in that one sample altogether. However, *when pooled together*, the reports of the four experts showed that *all* of the contested species were identified by at least two experts straight up to the K-T layer (and not above it), proving beyond a reasonable doubt that these extinctions all took place at that specific, razor-sharp level.[8]

Dinosaur, plankton and ammonite fossils all serve to illustrate a point made years earlier by micropaleontologists Philip Signor and Jere Lipps of the University of California: rare species always seem to fade away before their real level of extinction, simply because there are not enough fossils to guarantee that some will be found up to that level. Not to mention that the abruptness of a mass extinction also depends on how closely you are willing to look at it . . .

[8] Although a majority of species (70 %) go extinct at the K-T boundary, one should also note that another 30 % do cross the boundary, in decimated numbers but still without going extinct: these are called 'stragglers' or 'survivors'.

The shaping of a controversy

At this point in our story, it is worth reflecting upon the dichotomy that split the scientific community into such opposing camps, and looking into some of the mechanisms that steer the evolution of new ideas in science.

As an undercurrent to the whole K-T controversy, we already mentioned the reigning influence of gradualism in the Earth sciences. This paradigm holds that the Earth underwent only small and progressive changes over time, with no need for catastrophes outside human experience, and especially not from space. This attitude was colorfully illustrated in a 1985 article in the *New York Times*, stating that 'astronomers should leave to astrologers the task of seeking the cause of earthly events in the stars.'[9]

In order to better understand this belittling of impacts, one must stress again how little was known and taught of this geological process until recently. Most geologists and paleontologists trained in the fifties and sixties – which included those who later found themselves in the front lines of the K-T debate – had scarcely heard of impacts during their training, except as of an antediluvian process that played some role in the early days of the Solar System, four billion years ago, but had little to no influence today.

It was only during the sixties that NASA, the U.S. Geological Survey and North American universities began to map the Moon in preparation for the Apollo landings, bringing impact craters into the limelight, and driving geologists to search for their equivalents on Earth. Robert Dietz, Eugene Shoemaker, Bevan French and Frank Short in the United States, and M.R. Dence and Richard Grieve in Canada spearheaded the search for impact structures on the American continent and abroad. The crater hunt spread to Europe with the identification of the Ries Crater in Germany in 1967, closely followed by the discovery of the Rochechouart astrobleme in France.

But this knowledge was shared, even in the late seventies, by

[9] Cited in *The Mass-Extinction Debates: How Science Works in a Crisis* (p. 130), a remarkable collection of essays on the K-T controversy, compiled by science historian William Glenn (Stanford University Press, 1994).

only a few hundred scholars in the Americas, and fewer yet in Europe. This had a direct bearing on the positions that were taken on the issue.

One 'snapshot' of the beliefs and reactions of the scientific community to the impact hypothesis was taken in the Spring of 1984 by Antoni Hoffman of the Polish Academy of Sciences and Matthew Nitecki of Chicago's Field Museum of Natural History. The two scholars sent a questionary to several hundred scientists in North America and Europe, at a time when the iridium anomaly had been known for four years, but shocked quartz had not yet been reported.

Published in the December 1985 issue of *Geology*, the poll was answered by some 172 paleontologists and 82 geophysicists from the USA and Canada, 118 paleontologists from the United Kingdom, 113 paleontologists from Germany, 122 geologists from Poland, and 20 geologists from the USSR. Several trends clearly stood out.

Geologists from Eastern Europe felt little concerned about the issue (they had not been exposed to it to any great extent), while German and British paleontologists showed some interest in the developing controversy but rejected for the most part the cosmic hypothesis.

North American paleontologists and geophysicists were the only group where the K-T impact was accepted by a majority of votes, but there was an important split. American geophysicists accepted that an impact both occured *and* caused the mass extinction, whereas most paleontologists who went along with the idea of an impact did not necessarily believe it was the cause of the mass extinction.

This schism was confirmed the following year in the *New York Times*, in an article entitled *Dinosaur experts resist meteor extinction idea* (October 29, 1985). The article reported a poll, conducted at a meeting of the Society of Vertebrate Paleontology, in which only 4% of the polled paleontologists thought an extraterrestrial object had caused the demise of the dinosaurs, although 90% were willing to concede that an impact had occurred at the time. The main objection of paleontologists, again, was that dinosaurs had not disappeared abruptly, and that an abrupt impact was therefore not to blame.

Figure 3.4. Gradualists and catastrophists met many times in the field to discuss the evidence at the K-T boundary. Geologist Ken Hsü of Zurich University, one of the cofounders of the impact theory, examines a tsunami deposit at the Mimbral K-T site, Mexico. (*Photograph: by the author.*)

War of the worlds

Overall, the readiness of North American scientists to accept the cosmic hypothesis was certainly due to their greater familiarity with impact processes, and also to the fact that the founding article by Alvarez *et al.* was published in an American journal, *Science*, that was read principally in North America.

The poll also showed that geophysicists and geochemists accepted more readily the evidence available at the time than did classically trained geologists. In fact, most geologists did not truly embrace the impact theory until a few months after the Hoffman and Nitecki survey, when Bruce Bohor published his discovery of shocked quartz. This clue was much more understandable and acceptable to the majority of geologists than were obscure iridium levels and isotopic ratios.

One can wonder why so much time elapsed – a full four years – between the initial articles of Alvarez and others in 1980, and the discovery of shocked quartz in 1984. In fact, Bruce Bohor did ask for a grant to look for shocked quartz as early as 1981, but his request was turned down, so convinced were the grantors at the time that the cosmic hypothesis was not worthy of serious study. Bohor's second grant request was also turned down the following year. It was only in 1983 that the geologist set out on his own to look for shocked quartz, without the help of any grant, and met with success.

But it was in paleontology that the hostility toward the cosmic hypothesis ran deepest. Only micropaleontologists – specialized in fossils of plankton, pollen and other diminutive biota – were open minded about the issue, because they were familiar with sharp transitions in the fossil record, and paid attention to geochemical clues. When it came to larger animals (vertebrates and especially dinosaurs), the dogma of gradualism was seldom questioned and abrupt extinctions continued to be disregarded.

One anecdote is particularly telling in this respect. Canadian paleontologist Dale Russell, who in the late seventies was one of the few dinosaur experts who believed in their abrupt extinction, focused on the relevance of the K-T clay at about the same time as the Alvarez team. He held in his possession a fine sample of K-T clay from New Zealand, which he tried to analyze in Canada in 1979 to look for atomic anomalies (at the time, Russell suspected that a supernova explosion might have caused the mass extinction, and 'planted' rare atoms in the clay). But there was not much of a response to his solicitations for laboratory expertise, so odd was his request at the time. After all, didn't everyone know that the extinction of the dinosaurs had been gradual?

As it turned out, when the Alvarez paper was published in 1980, Dale Russell enthusiastically sent his New Zealand sample to Berkeley, since no one in Canada seemed in a hurry to analyze it. Sure enough, the Berkeley team discovered a strong iridium anomaly in the New Zealand clay and published the results the following year.[10]

The view from France

The opposition of gradualists to the impact theory was especially acute in France. Because I had the opportunity to follow it from the sidelines, this is my view of how the debate was shaped in one scientific and cultural microcosm.

In France, the opposition to catastrophism might be surprising when one recalls that it was a Frenchman – Baron Georges Cuvier – who championed the catastrophist theory in the early nineteenth century. But, over the years, the theories of Cuvier had been entirely dismissed. At the French Museum of Natural History, chairman and dinosaur expert Philippe Taquet has positioned himself today in favor of gradual extinctions, as has the director of vertebrate paleontology, Léon Ginsbourg, who believes that it was a progressive marine regression (a sea level fall) that caused the K-T mass extinction.

When the cosmic theory broke out in the early eighties, the attitude among French paleontologists was thus overwhelmingly

[10] Ironically, Dale Russell was then criticized by some of his colleagues for not having done the analysis in a Canadian lab!

critical. One should add that the cultural context did little to help. There was a general mood of defiance to ideas that came from North America. Moreover, there were only a couple of impact experts in France, and they were not high enough in the scientific pecking order to dare stick out their necks and take sides with the Alvarez theory.

The lopsided situation was aggravated in the mid-eighties when geophysicist Vincent Courtillot proposed the Deccan volcanic eruptions as the main culprit of the mass extinction. This view found much resonance in the French press since Vincent Courtillot was not only a renowned and talented scientist, but also one of the editorial advisors to the main scientific journal in France, *La Recherche*.

In the academic world, Courtillot's prominent political role – he was notably in charge of reforming science departments in the universities – might also have discouraged many a scholar from overtly disagreeing with him.

To the credit of the French research and grant system, however, much important work on the K-T boundary was funded even if it supported a cosmic model, as testified by the pioneering work of Robert Rocchia, Eric Robin and others.

As for the popular science press, it was more open minded than the scientific establishment and covered both sides of the debate, especially magazines familiar with space research and astronomy, like the monthly *Ciel et Espace*. Steven Spielberg's *Jurassic Park* (1994) also did much to heighten the public's interest in dinosaurs and their extinction. But overall, the opposition to the cosmic theory remained strong, because of the control exerted by senior scientists over the media.

An enlightening example is offered by a TV program on dinosaurs broadcast in the fall of 1994 by cultural channel ARTE. The segment on the extinction of the dinosaurs was divided into three parts. The first featured the volcanist theory;[11] the second presented

[11] The visual appeal of eruptions did a lot to promote the volcanist theory on television. Because the French TV viewership is notoriously fond of volcanoes, and because alternatively there was no spectacular footage of impacts (only rare and expensive animations), television producers were prone to highlight volcanic eruptions in their coverage of the debate. One exception was a science program on channel M6, which managed to portray the impact research of Robert Rocchia and Eric Robin in a visual and entertaining form.

Figure 3.5. French scientists approached the K-T mystery with widely different viewpoints. While Robert Rocchia (left) and his partners of the CEA/CNRS lab helped to gather evidence of an impact, Vincent Courtillot (right) and the IPG Institute of Paris pressed instead for drawn-out volcanic eruptions and gradual extinctions at the K-T boundary. (*Photograph: by the author.*)

the impact theory; and the third sent the first two back to back, concluding with Léon Ginsbourg's theory of a sea level drop as the probable cause of the mass extinction. The impression from the ARTE program was that no conclusion could be reached on the matter.

At the risk of getting ahead of our story, it is interesting to note how and when the impact theory was 'officially' accepted in France. The first, French edition of this book came out in the Spring of 1996. Revitalized by a shake-up of its editorial board, science magazine *La Recherche* finally published a special dossier on the impact evidence in December of 1996. Written by top European specialists in the field (Smit, Rocchia, etc.), the dossier ended with only one dissenting article, not by Frenchman Vincent Courtillot (who was apparently not consulted on this one), but by American paleontologist David Archibald.

Equally interesting was the reaction to this special issue. In the daily newspaper *Le Monde*, a follow-up article was entitled 'Volcanic eruptions in India and a giant meteorite caused the demise of the dinosaurs'. The volcanic theory was simply not going to die out. Although the article did present the impact theory in detail, the one sentence blown up and highlighted in its layout had only to do with volcanic eruptions.

Strong differences in opinion remain to this day in France, but exchanges remain cordial. As Dale Russell noted of his discussions with French colleagues, 'personal tolerance in the face of scientific disagreement may be a wonderful characteristic of French scientists'.

A change of mind

Those scientists who vigorously opposed the impact theory in the eighties and early nineties are now adjusting to the evidence. A few are still sticking to their guns, and continue to reject the cosmic hypothesis as pure fiction, calling it an example of 'pathological science'.

Others yield to the evidence in a stepwise fashion. At first they grudgingly accept the reality of an impact, but deny its role in the mass extinction. Later, they accept the reality of the lethal effects, but rate them as less important than contemporaneous 'gradual' causes of extinction, or accept them only as the final blow to an already doomed ecosystem.

Opinions were particularly rigid in the K-T debate, not only because two strong and opposing models – gradualism and catastrophism – were framing scientific thought, but also because the conflict of opinion was exacerbated by the unusually high level of public attention that put scientists on the spot. The growing popularity of the debate – which brought with it money and fame – acted as a deterrent to changes in opinion. In a heated, polarized debate of this kind, there was almost as much profit to be made in being wrong than in being right, since both sides got media attention and grant money.

Catastrophists and gradualists got caught up in this stand-off, and rarely attempted to strike any sort of compromise. The former could have let up a bit after proving the reality of an impact at the K-T boundary, and extended a friendly hand to the volcanists, acknowledging that the Deccan eruptions just *might* have fragilized the ecosystem, prior to the impact. This was seldom the case, however, mainly because impactists considered that the argumentation of their opponents had been fundamentally flawed from the start (namely regarding the iridium and shocked quartz). In the words of a major proponent of the impact theory, science is not a democracy. Scientists are either right or wrong, and there is no room for 'political' compromise.

Between the feuding camps, there were also those scientists who were long undecided and kept an open mind. Not only did these scientists get the most out of the debate, since they carefully weighed both sides of the issue, but they also made important contributions to the ongoing research, since they gathered and interpreted data with great objectivity.

An example was geochemist Robert Rocchia, who shared his expertise of iridium with volcanists and impactists alike. With the help of his team, he suggested experiments and lines of research that helped resolve a number of issues. For one, in the hope of better constraining the timeframe of the iridium deposition (the anomaly extends for several tens of centimeters above and below the K-T boundary),[12] Robert Rocchia and Eric Robin set out to independently measure the spread of meteoritic spinel crystals (see chapter 2). The sharp concentration of spinel crystals in the bottom few millimeters of the K-T clay brought independent proof of the brief nature of the event.

Incidentally, the spinel issue was the decisive argument that shaped Robert Rocchia's opinion in the matter, and illustrates how the mind adjusts in the face of conflicting information. Until that day in 1991, the Frenchman thought that the iridium 'smeared' on either side of the K-T clay suggested that iridium had started to rain from the sky thousands of years before the deposition of the K-T clay, and thus represented a drawn out event, more compatible with a shower of several comets (and stepwise extinctions) than with a single impact. When he saw the results of the spinel measurements – showing the meteoritic crystals all confined to a few millimeters of clay – Robert Rocchia left his laboratory perplexed by the findings. Over the next few hours he mulled over the evidence, attempting to reconcile the data with his quivering model of a drawn-out comet shower. It was not until nightfall that the scientist accepted the proof of one brief impact. Only then was he able to put his mind to rest and fall soundly asleep. It then took a few days for the idea to really sink in.[13]

Towards better communication

The K-T mystery gave rise to heated arguments and quarrels, but it also brought scientists closer together by stimulating interdisciplinary exchange.

[12] It now appears that this 'smear' of iridium on both sides of the K-T clay is probably due to several factors, including the time it took iridium to settle at the bottom of the ocean (prolonged by the input of more iridium from run-off water draining the continents), as well as the remobilization of the iridium in the sediment after its deposition, both through water percolation and 'bioturbation' by worms and other burrowing animals, all carrying the metal away from its initial, centimeter-thin layer of deposition.

[13] Another example of a change of mind is that of an American geologist, who initially did not believe that microspherules in the K-T layer were tektites from an impact. When he received convincing samples from Haiti, he was heard for a couple of days repeating to himself: 'My God, they really look like tektites!'

Figure 3.6. Dutch paleontologist Jan Smit, one of the cofounders of the impact theory, is interviewed by Mexican television. The K-T debate was covered extensively by the media, because of the public's unflagging interest in the dinosaurs and their mysterious extinction. (*Photograph: by the author.*)

In a little over ten years, over 2000 scientific articles on the K-T issue were published worldwide, in a variety of journals and languages. Because of the number of different fields involved – namely geology, paleontology, chemistry, physics, astronomy and biology – the K-T debate compelled specialists to explore areas of research that were unfamiliar to them.

In the process, they discovered new styles of rhetoric and communication. Each field of science has its own code of behavior and technical jargon. The latter is used as a tool of efficient communication, but also as an instrument of recognition and acceptance between 'initiated' members of a group. In the heat of the K-T debate, it was important for the players of each field to learn the vocabulary and 'etiquette' of other scientists in order to better understand their arguments and communicate their own.[14] Geologist Walter Alvarez humorously recalls that he caught the attention of physicists only when, instead of describing concentrations of iridium in the K-T clay as being 'one hundred times larger' than background, he used the expression 'two orders of magnitude greater', which was more convincing to physicists.

Finally, a word must be said about the positive role played by popular articles – and the media as a whole – in the tower of Babel of K-T research. Many scientists hard at work in their area of expertise had no time, and often not the proper training, to fully appreciate the discoveries made in neighboring (and not so neighboring) areas. They were kept abreast of recent developments by the editorials of research journals, articles in popular science magazines, and stories in the daily press, radio and television. Some scientists resented the intrusion of non-specialists on their turf – which came with its share of bias and oversimplification – but most welcomed this coverage as an important, constructive tool in the resolution of the K-T mystery.

[14] As early as 1981, Lee Hunt and Lee Silver invited colleagues from different fields in K-T research to a seminar at Snowbird, Utah, with the goal to share knowledge and improve communication between the disciplines. The experience was repeated with growing success at Snowbird in 1988, and in Houston in 1994.

4

In search of the crater

By the end of the eighties, there was clear evidence that the impact of a large-sized bolide had occurred at the K-T boundary, based on the geochemical and mineral clues found in the conspicuous layer of clay. But a major piece of evidence was still lacking: where was the giant crater itself, which the collision should have blasted somewhere on the Earth's surface?

The amount of iridium in the K-T clay, integrated over the surface of the globe, had allowed the Alvarez group to estimate the impactor to be on the order of ten kilometers in size, which implied that the resulting crater should be close to 200 km in diameter.[1]

One expected such a large crater to be immediately recognizable, but such was not the case. There were no large impact craters dated at 65 million years. Volcanists and other gradualists could go on challenging the cosmic hypothesis until an impact crater of the right size and the right age was found.

Supporters of the impact theory were clearly frustrated by the lack of an obvious crater, but far from surprised. They knew how difficult it would be to discover an impact scar, or, *astrobleme*[2] some 65 million years old on the Earth's surface, given the forces of erosion, burial, and tectonic destruction that are at work on our dynamic planet. In addition, if the impact had occurred at sea (two chances out of three, given the ratio of oceanic to continental area), then the crater would be particularly difficult to locate for several reasons. For one, the search for submarine craters had barely begun, plagued by limited data of sufficient resolution (the best geophysical surveys of the ocean floor were military, and hence classified).

[1] Experiments and computer modeling show that an impact crater on Earth is approximately 20 times larger than the impactor's diameter.

[2] 'Astrobleme' is the name given to an old impact crater on Earth, degraded by erosion.

Another reason was that close to shore, on the continental shelves, impact craters would be buried under large amounts of coastal sediment, further hindering their identification.

But the most adverse mechanism in the oceanic realm was plate tectonics, which relentlessly erases the impact record of the ocean floor by destroying it in subduction zones – deep trenches at plate boundaries, where the creeping oceanic plates bend down and vanish into the hot batter of the Earth's mantle. Impact craters stamped onto the ocean floor thus creep to their destruction at the rate of a few centimeters per year. Modest as it may seem, this pace signifies that about a quarter of the oceanic crust present at K-T time has since been destroyed: odds of finding a 65 million-year-old crater on the ocean floor were only three chances in four.

Down the K-T trail

The task of finding the crater was especially daunting because the K-T layer offered very few clues as to the location or direction of impact. Expectations that the clay may thicken towards its 'source', for instance, had been dashed when it was realized that the clay thickness was nearly the same all around the globe, and that its slight variations were due more to local sedimentation patterns than to any distance-related function.

There was one slight hope, however. In North America, at the foot of the Rocky Mountains, the K-T layer did look as if it was thicker than average (two centimeters rather than one), as well as dividable into two separate parts: the iridium-laced, spinel-rich 'traditional' clay; and a coarser section above it, bearing the shocked quartz grains – grains that appeared to be larger than at other K-T sites. Several scientists interpreted this thickened layer to mean that the crater was on the American continent or nearby.

The K-T sleuths also tried to find mineralogical and geochemical clues in the clay that would shed light on the type of terrain pulverized by the impact, namely if it was continental (silica-rich) or oceanic (calcium and magnesium rich). At first glance, however,

Figure 4.1. Artist's view of a large impact crater at sea, carved in the shallow continental platform. Although the crater's ramparts are initially gouged by erosion, sea-level rises can drown the feature and cap it with a blanket of sediment, ensuring its long term preservation. (*Photograph: courtesy of William K. Hartmann (© William K. Hartmann).*)

the issue was confused: some microspherules in the K-T layer were apparently basaltic, suggesting an impact on the oceanic crust, while on the other hand the abundance of siliceous minerals – shocked quartz and also sanidine, a variety of feldspar – pointed to a continental source.

One way to interpret the continental quartz and the would-be oceanic spherules was to imagine that the impact took place on the border of continental and oceanic terrain, somewhere on the continental shelf: this was a clever compromise but the odds of such a strike were low, given the small area of this type of terrain at the Earth's surface.

Another possibility was that the K-T crisis could have resulted from multiple impacts. A salvo of cometary or asteroidal fragments could have struck both oceanic and continental terrain simultaneously, leading to the mix observed in the K-T clay.

Regardless of one's preference, it had become clear by the end of the eighties that one crater at least was to be found on a continent (or continental shelf), because of the K-T quartz. Reassured that the evidence was not lost at sea, crater hunters renewed their efforts to find the great K-T crater.

A primer of impact craters

Although they didn't know exactly where to look, crater experts knew what features to look for. Since the pioneering days of impact studies in the late fifties, planetary geologists had amassed a substantial expertise with respect to the mineralogy and structure of impact scars on Earth. Several new sites were identified each year. From only half a dozen astroblemes recognized on Earth in the mid-sixties, the tally had reached one hundred impact craters in the mid-eighties.[3] See table 4.1.

Based on this experience, scientists expected the K-T crater would be a broad circular structure, not unlike a volcanic caldera. Astroblemes resemble volcanic calderas at first glance. Both can contain igneous rock – ash and lava flows in a caldera; impact melt in an impact crater. But there are ways to tell the difference. Whereas volcanic rocks display chemical trends that show them to be derived from the Earth's mantle at depth, impact melts show a more exotic chemistry that reflects the composition of the local target rock. Their texture and mineralogy point to very high temperatures of formation, much above the range of ordinary lavas. Some impactites are known to include high-temperature forms of silica, such as *lechatelierite*[4] that crystallizes above 1500 °C, well out of the temperature range of volcanic magma.

These *impactites*, as they are called, often contain the tell-tale signature of the volatilized cosmic body itself, in the form of unusually high contents of nickel, platinum, iridium and other siderophile elements in meteoritic proportions.

[3] In addition, geologists were now familiar with impact features on other planets: astronauts had studied and sampled impact craters on the Moon, not to mention the impact craters of Mars, Venus, Mercury and the moons of the giant planets, imaged by a number of automatic space probes.

[4] This unusual form of vitreous quartz was first identified by French mineralogist Le Chatelier and named after him.

Table 4.1 *The twenty largest astroblemes identified on Earth as of 1997 (ages are indicated in millions of years). In the 1980s, Chicxulub crater had not yet been discovered.*

Name	Location (Lat, Long)	Diameter (km)	Age (m.y.)
Vredefort	South Africa (27 °S, 27 °E)	300	2023
Sudbury	Ontario, Canada (47 °N, 81 °W)	250	1850
Chicxulub	Yucatan, Mexico (21 °N, 90 °W)	180	65
Lake Acraman	Australia	160	580
Manicouagan	Quebec, Canada (51 °N, 69 °W)	100	214
Popigai	Russia (71 °N, 111 °E)	100	35
Chesapeake Bay	Delaware, USA (37 °N, 76 °W)	85	35
Puchezh-Katunki	Russia (57 °N, 44 °E)	80	175
Morokweng	South Africa (26 °S, 23 °E)	70	145
Kara	Russia (69 °N, 65 °E)	[a]65	57
Beaver Head	Montana, USA	60	>600
Tookoonooka	Australia	55	128
Siljan	Sweden (61 °N, 15 °E)	52	368
Charlevoix	Quebec, Canada (48 °N, 70 °W)	46	360
Montagnais	Nova Scotia (43 °N, 64 °W)	45	50
Mjølnir	Barents sea (72 °N, 30 °E)	40	144
Araguinha Dome	Brazil (17 °S, 53 °W)	40	247
Carswell	Saskatch., Canada (58 °N, 109 °W)	37	117
Manson	Iowa, USA (42 °N, 95 °W)	35	74
Clearwater Lake West	Quebec, Canada (56 °N, 74 °W)	[a]36	290

Note: [a] Main astrobleme of a pair (multiple impact). Kara has a twin structure, Ust Kara, 25 km in diameter. Clearwater Lake West also has a twin structure to the East, 26 km in diameter.

Astroblemes are also remarkable for their signs of high-pressure deformation. At a microscopic level, these range from shock defects in crystals to completely new mineral forms that occur only at very high pressures, such as *coesite* and *stishovite*.[5]

[5] Coesite and stishovite form at pressures above 4 GPa and 13 GPa respectively (40 000 and 130 000 atmospheres).

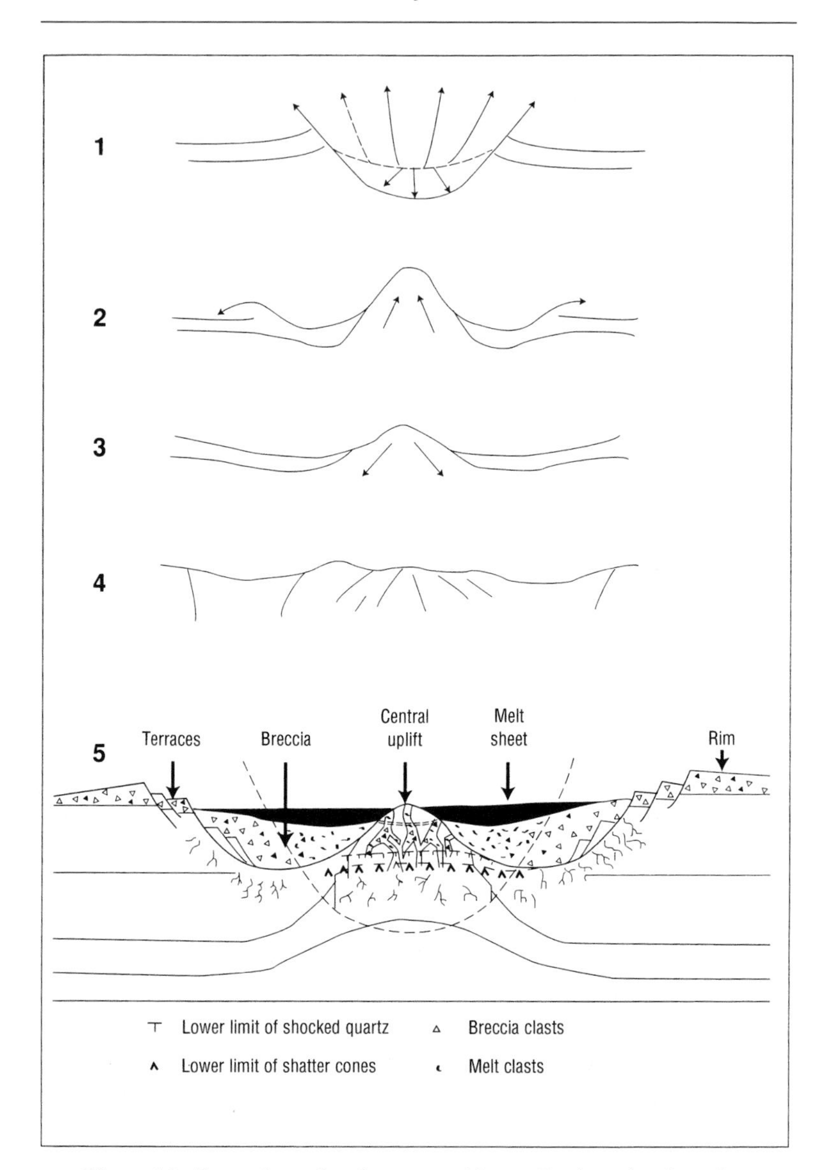

Figure 4.2. Formation of a large astrobleme. During the first few seconds of impact, the upper layers of the target are ejected, and a compression wave propagates through the ground (1). In the wake of the compression wave, the crust relaxes and rises in a central rebound (2), exposing deep crustal rock as a central peak or plateau (3). As the crater floor settles, ring faults propagate around its periphery and

On a broader scale, shock pressures create features that are visible to the naked eye, such as *shatter cones* – stress patterns that propagate through fine-grained rocks on the site of the impact and lace them with small cones pointing in the direction of ground zero. Other evidence of impact are the disturbed target rock, and notably the *impact breccias* which consists of clasts of broken-up bedrock, fused together in a glassy matrix.

The crater takes shape

A large impact creates a nearly circular depression. Hitting the planet at velocities of tens of kilometers per second, large projectiles plow through hundreds of meters of target rock in a fraction of a second, propagating before them intense shock waves that compress and then decompress the rock within the fast-growing cavity.

In the case of projectiles larger than one kilometer in size, which create craters larger than 20 km on the ground, the vigorous blow-out literally punches a hole in the atmosphere, a 'vacuum tunnel' through which the unimpeded ejecta soars skyward before raining back far and wide around the crater. In the center of this fiery jet, the highest-temperature ejecta is propelled high above the stratosphere, spreading outward as a 'fireball' of vaporized material: in the case of the K-T impact, this fireball is believed to be the source of the iridium-rich clay deposited worldwide.

As for the coarser ejecta lofted on the periphery of the blow-out, most of it falls back within one or two crater diameters around ground zero. Known as the 'proximal ejecta', this blanket of rubble reaches several hundred times the mass of the striking projectile.

At this initial stage of crater formation, one can imagine a gaping,

Caption for fig. 4.2 *cont.* downdrop concentric terraces (4 and 5). A well-preserved astrobleme typically displays shattered basement rock, a fill of broken-up breccia (the ejecta), and pods of impact melt. (*Modified from P. Hodge,* Meteorite craters and impact structures of the Earth, *1994, Cambridge University Press; and V. L. Sharpton and R.A.F. Grieve, Geological Society of America Special Paper 247, 308, 1990.*)

Figure 4.3. Impact crater on the Moon, as viewed from an Apollo spacecraft. Because it has no atmosphere and hence little erosion, the Moon has preserved a nearly full record of its cosmic impacts over the past four billion years. Large impact craters are characterized by ring terraces, due to the slumping of the original cavity. A central peak, due to crustal rebound, is present in the crater center. (*Photograph: NASA.*)

deep bowl with glowing impact melt cascading down the steep, ephemeral slopes. Titanic landslides quickly enlarge the unstable pit, sliding material in from the sides along arcuate, spoon-like faults. After a few minutes of havoc, the crater reaches its adjusted configuration, its settling slopes showered with the last blocks of ejecta falling from the sky.

The shape of an impact crater is strongly size-dependent. On Earth, craters are simple and bowl-shaped up to diameters of about four kilometers. At larger sizes, a prominent peak stands out in the center of the structure, and there are terraces at the periphery that step down to the crater floor (see figure 4.2).

On the Moon these complex craters are barely touched by erosion and can be studied in great detail (see figure 4.3). Central peaks were first observed on the Moon, where they were initially interpreted to be mere landslides, converging from the slopes of a crater to build a mound in its center. Now it is known that these central peaks mark the elastic rebound of the compressed basement at ground zero, the deep rock lifting skyward and 'freezing' in place.

On Earth, central peaks are found at astroblemes up to a dozen kilometers in diameter. Larger impacts involve the uplift of an entire plateau or ring. One example is Gosses Bluff in Australia, 22 km in diameter, which boasts a central ring 5 km wide (see figures 4.4 and 4.5). At Gosses Bluff, geologists estimate that the uplifted rock surged from a depth of at least two thousand meters. Since the central rebound is capped by a layer of fallback breccia, the deep crustal rock took only a few minutes to thrust its way to the surface, in time to intercept the ejecta raining from the sky. Gosses Bluff is therefore an outstanding example of the brutal rapidity at which impact processes take place. Not only are depressions several kilometers deep shaped in a matter of seconds, but mountain peaks can surge thousands of meters in a matter of minutes.

When it comes to even larger craters, structures become even more complex. The central uplift tends to spread out into a lower, more subdued relief, probably because the impact energies are so high that the basement rock in the center of giant astroblemes is nearly liquefied and unable to sustain heavy loads. Spreading outwards, the rebound wave 'freezes' at some distance from the center as a ring of hills. These ring basins are prominent on the Moon, but have yet to be discovered on Earth.

What shape, then, would the K-T crater have? Since the iridium content of the clay and other clues point to a crater 200 km in diameter, such a crater would obviously be of the complex kind. Its central uplift would be a ring or plateau about 40 km wide, surrounded by a large annular trough filled with fallback breccia and impact melt rocks. Peripheral terraces would complete the picture, stepping up to a scalloped rim marking the outer margin of the structure.

The list of candidates

In the mid-eighties, when the search for the crater picked up, there were two ways of proceeding. One was to consider that the K-T

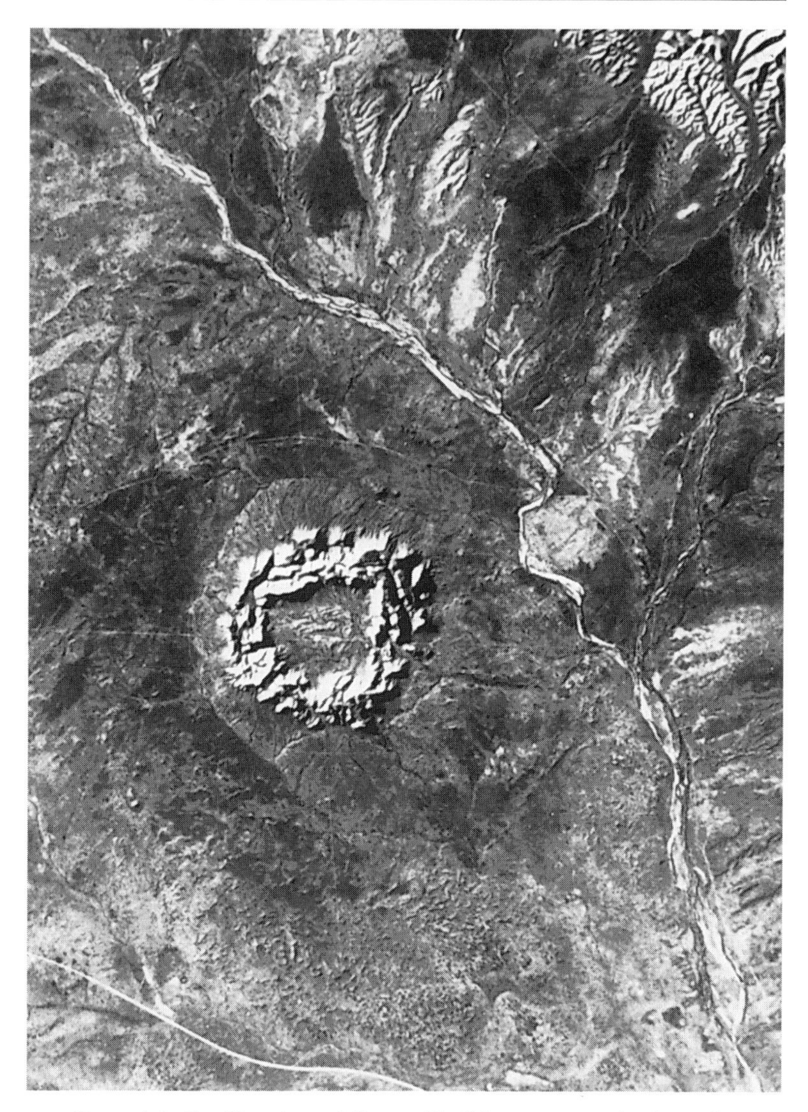

Figure 4.4. Satellite view of Gosses Bluff impact crater in Australia. The bright crown of relief, five kilometers in diameter, is but the central rebound of a much larger astrobleme, bounded by an outer ring of darker terrain. Gosses Bluff crater is 22 km in diameter, and its age is estimated at 144 million years, contemporaneous with the End-Jurassic mass extinction (see chapter 7). (*Photograph: NASA, courtesy of Richard A.F. Grieve, Natural Resources Canada.*)

Figure 4.5. Aerial view of the Gosses Bluff central ring, caused by the rebound of compressed crustal basement following the impact. Since impact time, the softer rock strata in the middle of the ring have been eroded away, but the ring itself still shows layers of fall back ejecta overlying the uplifted crustal rock. (*Photograph: courtesy of Richard A.F. Grieve, Natural Resources Canada.*)

crater was one of the astroblemes already known, and pick it off the list. The other possibility was that the crater was still concealed and awaited to be unearthed.[6]

The first approach did not look promising. Of the one hundred astroblemes known in the mid-eighties, none stood out as having both the right size and the right age (see table 4.1). There were only two large astroblemes of K-T size – Sudbury in Canada (200 km) and Vredefort in South Africa (140 km)[7] – and both were

[6] One also had to keep in mind that a single crater was not the only way to explain the volume of ejecta spread around the globe: several simultaneous, smaller impacts could also do the trick.

[7] The Vredefort and Sudbury astroblemes are now believed to be respectively 300 km and 250 km in width.

over 1.8 billion years old, thirty times the age of the K-T event!

That there were no giant craters younger than 1.8 billion years was perplexing in its own right. Probabilities based on lunar crater counts and other considerations hinted that K-T sized bodies should have hit the planet every 100 million years or so. Even if two impacts out of three occurred at sea (their craters eventually destroyed in subduction zones), there still should be one large crater formed on dry land every 300 million years.

Hence, since the days of Vredefort and Sudbury, half a dozen craters over 150 km in size should have collected on the continents. Where were they? Arguably the forces of erosion – and burial by sediments – must have been tremendously effective to conceal these large craters. The discrepancy also indicated that chances of finding any one crater were low – one in six or worse.

Be it as it may, no giant crater posted itself as the obvious source of the K-T clay. No medium-sized crater did the trick either. The ubiquitous Manicouagan astrobleme in Quebec – an impressive 100 km wide – was over 200 million years old, which cleared it of the K-T massacre, as was cleared the large (80 km) Puchezh crater in Russia, also dated at over 200 million years. Of the younger astroblemes, only Popigai in Russia boasted a large size (100 km) but this time it was too young: 35 million years.

We shall return to all these craters in chapter 7, and check if they were not involved in other mass extinctions of the biosphere. But there were clearly no large astroblemes of Late Cretaceous age.

Among the smaller craters, however, several did call out for attention. As early as 1984, encouraged by the discovery of shocked quartz in the K-T layers of Montana and New Mexico, geologist Bevan French suggested two North American impact sites: Manson in Iowa and Sierra Madera in Texas. The latter was quickly ruled out: besides its small size (16 km), Sierra Madera was soon found to be much older than the K-T boundary.

Manson crater, on the other hand, was a promising candidate: over 35 km in size, it appeared to sit in target rock of roughly the right age that was rich in quartz, and it was less than 2000 km away from the North American K-T sites that contained the largest shocked minerals.

Figure 4.6. Map of impact structures identified on Earth, as of 1997. The highest concentration of astroblemes occurs on old and stable continental basement rocks that have registered impacts over hundreds of millions of years, notably in North America, Scandinavia and Australia. (*Courtesy of Richard A.F. Grieve, Natural Resources Canada.*)

Manson, the main suspect

Manson was quite an interesting crater in its own right. Geologists had long been intrigued by the unusual stratigraphy of the site, in the middle of Iowa's corn belt. Under a hundred meters of glacial till, drillings conducted shortly after World War II had revealed a circular structure over 35 km in diameter, in which the sedimentary strata were thoroughly disrupted. Around the anomalous structure, an outer ring extended an extra 10 km and was missing its uppermost strata, as if they had been uplifted and stripped by erosion. Last but not least, geophysical surveys showed that the crystalline basement was uplifted by several thousand meters in the center of the structure, almost reaching to the surface.

By 1953, drill cores had recovered samples of the central uplift, and in 1959, impact specialist Robert Dietz visited the site to assess the evidence and propose an impact origin for the Manson structure.

The verdict was confirmed in 1966 by Frank Short, who discovered shocked quartz in cores of the basement rock.

Manson was a good example of a complex crater with a central peak. The crystalline uplift was 12 kilometers wide, with a vertical displacement of about 4000 meters. Gravity anomalies measured by Allan Holtzman in 1970 further refined the picture by pointing to a thick ring of low-density breccia surrounding the central peak. Drilling on the outskirts of the crater showed the rim sediments to be flipped inside out – the older strata overlying the younger – as is often the case in proximal ejecta blankets.

In the mid-eighties, when the K-T sleuths focused their attention on Manson crater, what interested them most was its age. It was common knowledge that local sediments older than the Upper Cretaceous Dakota formation had been disrupted by the impact, whereas younger sediments smoothly covered the structure, hence postdating the event. At first glance, then, Manson did appear to be End-Cretaceous in age. But only radiochronology age measurements, performed on samples from the crater, would settle the issue.

The first age estimates, published in 1986, were drawn from shocked feldspar crystals, using the potassium/argon method: an age of 70 million years was obtained, with a margin of error of plus or minus several million years – too large an uncertainty to draw any conclusions.

Two years later, in 1988, finer age measurements were obtained with the new argon–argon technique, which was proving itself as the most precise dating tool of the eighties and nineties. With this new technique, the shocked feldspars indicated an age of 65.7 ±1 million years, and Manson crater became a major suspect of the K-T mass extinction.

The Russian candidate

But Manson was not alone. Russia also harbored a promising impact site on its Arctic shoreline, near the estuary of the Ob River: Kara crater.

Remarkably concealed, the Kara astrobleme was nearly impossible to detect on aerial and satellite imagery. It was its scattered ejecta that had caught the attention of Russian geologists at the turn of the century. The breccia was mistaken at first for volcanic tuff, nested in some giant caldera. In the mid-seventies, however, impact specialists combed over the frozen tundra and identified the tell-tale signs of impact: shatter cones, impact glass and shocked quartz. The crater was buried for the most part under a thick cover of glacial till, but the many rivers that drained the tundra had eroded the structure in several places, exposing layers of breccia and impact melt. Geophysical surveys published in 1980 revealed that the impact basin was 65 km in diameter, with a buried central peak 10 km wide. In the basin trough – between peak and rim – the gravity measurements hinted at a sequence of breccia 1000 to 2000 meters thick.

The first age measurements of the Russian astrobleme, obtained in 1980, came out at 60 million years, but with a wide uncertainty of plus or minus 10 million years. The figure was encouraging enough to propel Kara to the top of the K-T suspects list, second only to Manson crater. In 1989, new analyses pinned down the age to approximately 66 million years – an even better fit.

Not only did Kara boast the right age for the K-T crisis, but it also was relatively large (65 km), and perhaps doubly so when clues hinted that it might be a double crater.

Double craters are caused by pairs of asteroids or comet fragments that travel together in space, and strike their target as a double hit. Double objects have been imaged by space probes, the prime example being asteroid Ida and its small companion Dactyl, photographed by the Galileo space probe in 1993. One famous pair of impact craters on Earth is the Clearwater Lakes of Quebec, blasted side by side in the Canadian shield: the western and eastern astroblemes are respectively 36 and 26 kilometers across (see figure 4.7).

Russia's Kara crater appears to be similar: to the east of the 65 km crater, a twin feature was identified in the mid-seventies and its size initially estimated at 25 km, but it was buried under the freezing waters of the sea of Kara, which considerably hindered its study. Known from only a few impact melt rocks exposed along

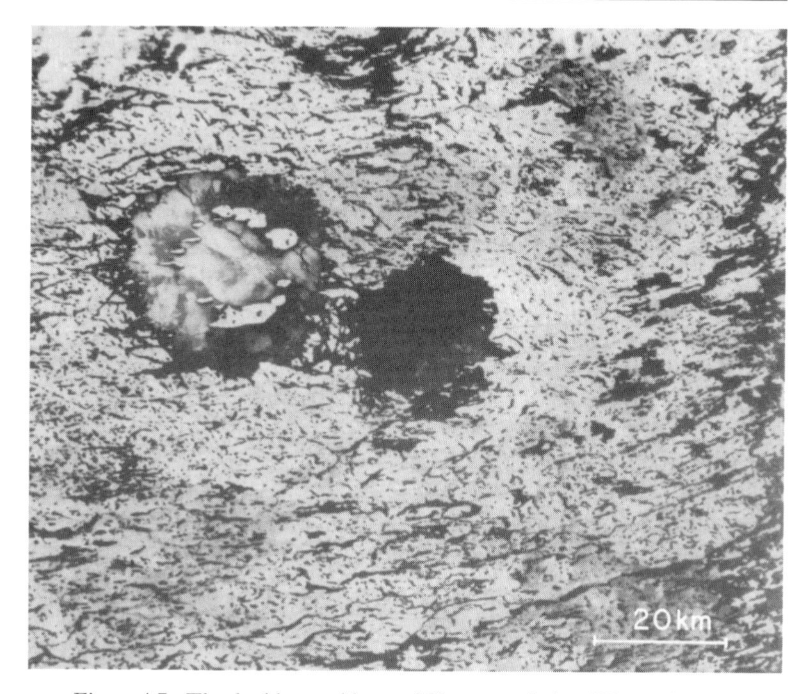

Figure 4.7. The double astrobleme of Clearwater Lakes (West and East), photographed from orbit. The simultaneous hit, 290 million years ago, is probably due to a couple of asteroids travelling together in space, or to fragments of a broken-up comet. The craters are 26 and 36 km in diameter, and the initial projectiles must have measured one and two kilometers in size. (*Photograph: NASA.*)

the coastline, this eastern crater – tentatively named Ust Kara – became the focus of much attention when the K-T sleuths zeroed in on the site.

Christian Koeberl and Virgil Sharpton in particular, working out of the Lunar and Planetary Institute in Houston, pored over satellite data in an attempt to better constrain the size of the eastern astrobleme. Satellite imagery *per se* was useless, since the crater was submarine, and most likely buried under sediment. On the other hand, gravimetric profiles collected by the Seasat and Geosat spacecrafts proved useful. They made it possible to search for buried structures, since buried masses affect the local gravity. In fact, the gravimetric profiles collected over the sea of Kara revealed two gravity peaks 70 km apart, which Koeberl and Sharpton interpreted as the uplifted rim on either side of the astrobleme.

If this interpretation was correct, then the eastern Kara crater was not 25 km but 70 km wide, making it even larger than its western counterpart. Together, Kara and Ust Kara would represent a substantial force of impact, and gain credibility as a major player in the K-T crisis.

Argon–argon alibis

Crater hunters were confident that they were well underway of indicting Manson and Kara in the great K-T massacre. Better yet, several geologists began to believe in a vast cosmic conspiracy, in which the Manson and Kara astroblemes were only part of a larger salvo of projectiles – a comet or asteroid which had split up in pieces before blasting the Earth.

The concept of a deadly salvo had been suggested by Rezanov as early as 1980. The Russian scientist remarked that the Kara craters, Kamensk crater in Ukraine (25 km in diameter and also believed to be of K-T age) and a small crater in Libya were aligned on a great circle spanning thirty degrees – a twelfth of the Earth's circumference.

Virgil Sharpton and Kevin Burke proposed a similar scenario in 1988, adding to the list their favorite suspect, Manson Crater, to form a great circle spanning eighty degrees around the globe. Since this great circle passed near the north pole, Sharpton and Burke suggested that other astroblemes might lie concealed at the bottom of the Arctic ocean, and signaled a 200-km-wide circular basin in the Mendeleiev abyssal plains.

The sudden flurry of suspects enchanted the supporters of the impact hypothesis, but the case was far from closed: none of the listed craters could be convicted beyond the shadow of a doubt. On the contrary, as time went by, the main suspects managed to consolidate their alibis: contrary to earlier evidence, they all turned out to be too old to account for the K-T massacre.

This realization came with improvements in the argon dating

technique, which provided better age estimates for both Kara and Manson Craters – ages that began to depart significantly from the magic number of 65 million years. The first suspect to be discharged, in 1990, was the Kara double astrobleme: its new age estimates fell between 73 and 76 million years, a full 10 million years older than the K-T event. Soon thereafter, Manson crater also posted an older age than previously believed: 74 million years, with a narrow margin of error (500 000 years) suffering no contest. To the great disappointment of its prosecutors, Manson Crater was cleared as well of the great dinosaur massacre.

With all the main suspects discharged, it became evident that the K-T astrobleme was still at bay, hiding somewhere on the Earth's surface . . .

5

The discovery of Chicxulub crater

By the late eighties, the K-T sleuths had come up against a dead end. No astrobleme seemed to fit – in terms of size and age – the End-Cretaceous catastrophe. Geologists needed to find new clues that would orient their search in the right direction.

Fortunately, the cosmic hypothesis provided several tests that made this possible. For example, it was obvious to some geologists that in the case of an oceanic impact, the shores closest to ground zero would have been swept by giant tidal waves or *tsunami*, leaving in their wake thick layers of disturbed sediment on the continental platform.

Geologists were on the lookout for such deposits, and in 1985 Dutch paleontologist Jan Smit signaled an unusual outcrop of K-T sediment in Texas, on the outskirts of the Gulf of Mexico. There, in the rock layers exposed by the erosion of the Brazos river, lay a bench of sandstone at exactly the K-T boundary, which Smit and coauthor Romein interpreted to be a tsunami deposit laid down by the impact.[1]

The banks of Brazos River

Although the tsunami interpretation was largely ignored at the time, a few geologists picked up on the news, including sedimentologist Joanne Bourgeois of Washington University.

[1] The unusual coarse sediment had been signaled as early as 1981 by Gartner *et al.* Its significance as a giant wave deposit was independently explored in the mid-1980s by Smit and Romein, and by Alan Hildebrand.

Figure 5.1. Around the Gulf of Mexico, unusual outcrops are found at the K-T boundary. Here at El Peñon, Mexico, a thick sandstone unit is interpreted to be a catastrophic tsunami deposit, laid down by the impact. It is capped by a fine clay displaying a wavy pattern, thought to mark the oscillation of the current as the tsunami wave sloshed back and forth across the continental platform. (*Photograph: by the author.*)

Bourgeois and her team pored over a number of outcrops at Brazos River and in the banks of its confluent stream, Darting Minnow Creek. They described the K-T boundary as consisting of unusually coarse sandstone, a few tens of centimeters thick, loaded with shell fragments, fossilized wood debris, fish teeth and large chunks of clay torn up from the local sea floor. In Cretaceous time, the Brazos site lay under 100 meters of water.

From the size of the clay blocks mixed up in the sandstone, Bourgeois *et al.* calculated that the bottom current at Brazos must have exceeded one meter per second – quite a large velocity for such deep waters. Further evidence of a tsunami came from a sheet of clay capping the sandstone, with wavy patterns that pointed to an oscillatory current – toward *and* away from shore. Such oscillatory currents are

Figure 5.2. Cross-section of the tsunami clay at the top of the El Peñon sand unit. The undulating lines are profiles of the ripple marks that built up in the clay as the current switched directions back and forth. (*Photograph: by the author.*)

typical of tsunami deposits, and are generated by the reflection of tidal waves off the continental platform (see figure 5.1).[2]

Retaining the model of an impact tsunami, Joanne Bourgeois and her co-workers attempted to estimate the amplitude of the wave that swept the Brazos site, and guess how far they were from the point of impact. They worked from the assumption that a powerful impact at sea would set off a wave as high as the water is deep at that point – a staggering wave height of 4000 to 5000 meters for a large impact in the middle of an ocean. The amplitude of the wave

[2] Before endorsing the tidal wave theory, Bourgeois and her coauthors ruled out other possible origins for the sandstone, such as turbidity currents and coastal storms. Turbidity currents, which are flows of dense, mud-laden water down a submarine slope, would have shown current in one direction only (downslope), and not the flip-flop pattern exhibited in the Brazos clay.

As for coastal storms, they could not possibly have displaced meter-sized blocks of clay over such distances and to such depth.

then decreases regularly with distance away from ground zero. Hence, if one knows the tsunami wave height on a site, one can deduce the distance to the point of impact.

From the estimated velocity of the sea-bottom current at Brazos – at least one meter per second, according to the size of the transported clay blocks – Bourgeois and her coauthors estimated that the wave amplitude at Brazos was at least 100 meters. This meant that the point of impact was less than 5000 kilometers away, somewhere in the Gulf of Mexico, the Caribbean or the western Atlantic.[3]

The hills of Haiti

The sandstone outcrops of Brazos River were not the only evidence of an impact near the shores of North America. In 1986, a Polish geologist, Andrzej Pszczolkowski, signaled thick layers of coarse, sandy debris in Cuba, at or near the K-T boundary. He, too, suggested that they might be the mark of an impact. The article, written in Spanish and published in a Polish journal, did not get much attention at the time.[4]

But it was in Haiti that the most important clues were uncovered. A report by Haitian geologist Florentin Maurasse described unusual sediments on the southern tip of the island, at the Cretaceous–Tertiary boundary. Described as 'volcanogenic', this deposit caught the attention of a young Canadian geologist, Alan Hildebrand (see figure 5.3).

A graduate student in planetary science at the University of Arizona, Alan Hildebrand had focused his research on what he believed was the impact wave deposit from the K-T crater, all along the Gulf coast. He looked at cores of K-T sediment drilled from the sea bottom, as well as outcrops on shore like Brazos River, before turning his attention to the Haitian deposits described by Maurasse.

After trying in vain to have samples sent to him, Hildebrand flew at his own expense to Miami to meet Maurasse at Florida's

[3] Assuming that the starting wave height was 5000 meters.

[4] The Cuba sandstones are still poorly understood and might be unrelated to the impact.

International University. Upon seeing the Haiti samples, Hildebrand was convinced that these were not volcanic deposits but impact ejecta, complete with altered tektites, and resolved to go on location and study the outcrop in person, with the help of fellow student David Kring.

When they reached the Massif de la Selle on the southern coast of Haiti, the two scientists found the outcrop as it was described in Maurasse's report. The unusual stratum that ran across the hillside was half a meter thick, its olive-green luster contrasting with the yellowish-brown limestone directly above and below it. To the naked eye, this K-T stratum appeared to be composed of countless spherules and other rounded particles that looked like impact tektites (see figure 5.4).

Hildebrand was all the more convinced that these were impact ejecta when he also found abundant shocked quartz in the spherule bed – many of the grains reaching a centimeter in size, even larger than in North America. In addition, the sandstone was capped by a fine clay a few millimeters thick, rich in iridium, which he interpreted to be the classic K-T 'fireball', settling last and draping the blanket of ejecta.

The thickness of the spherule bed in Haiti convinced the scholar that he was closing in on ground zero. From scaling equations that related ejecta thickness to distance from the source, half a meter of fall back corresponded to a distance to the crater of approximately 1000 kilometers. Clue after clue, the K-T sleuths were closing in on ground zero.

The Caribbean revisited

In the Late Cretaceous, the region between the Americas offered a somewhat different picture than today. The plains of the Gulf of Mexico were flooded by higher waters, up to the foothills of the Sierra Madre Orientale. To the east, the shallow Yucatan platform was flooded under almost a hundred meters of water. As for the sediments that would later form the islands of the Greater Antilles – like Cuba and Haiti – they were part of the deep sea floor and

Figure 5.3. Canadian geologist Alan R. Hildebrand at work on the K-T boundary. The impact deposit thickens from two centimeters in continental North America to 50 centimeters in Haiti, pointing to a target site near the Yucatan peninsula. (*Photograph: courtesy of Alan R. Hildebrand.*)

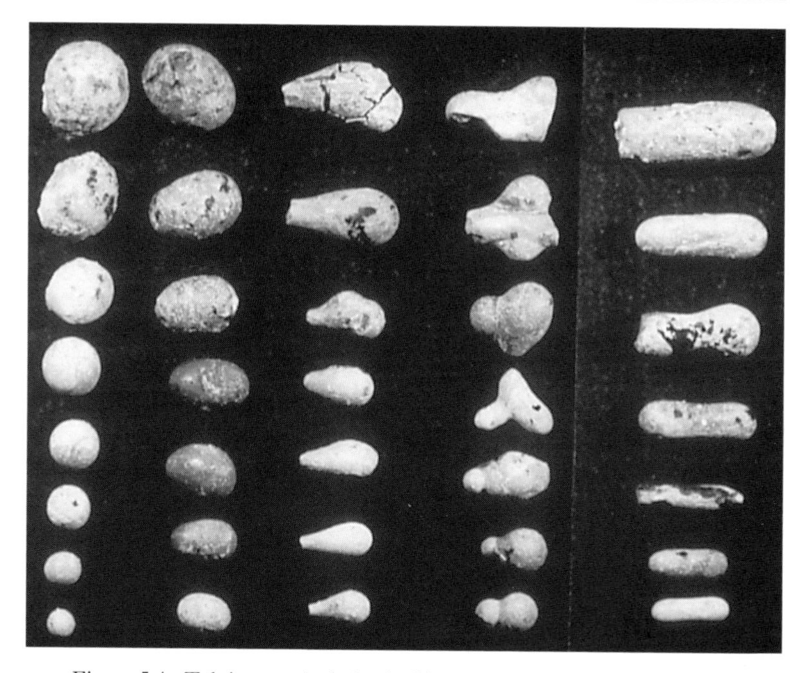

Figure 5.4. Tektites are 'splashes' of impact melt that take on aerodynamic shapes as they spin through the Earth's atmosphere. Tektites up to a centimeter in size are abundant in Haiti's K-T layer, betraying the proximity of the source crater. (*Photograph: by Alan R. Hildebrand.*)

had yet to be lifted by fault movements up to the surface. At the time of the K-T event, the limestones of Haiti lay in fact at depths of over 2000 meters, as indicated by the abyssal microfossils they contain. To further complicate the picture, the sea-floor segment that would later become Haiti lay a thousand kilometers southwest of its present location, closer to Yucatan, on the margin of an expanding Caribbean basin (see figure 5.5).

It is in this complex tectonic framework that Alan Hildebrand set out to look for the evasive impact crater, in a 1000 kilometer radius around Haiti's 'paleoposition' – the position that it occupied in K-T time. As he pored over maps, he first focused on an intriguing semi-circular structure on the outskirts of the Columbian Basin. Almost 300 km wide and buried under 2000 meters of sediment, it seemed to feature some central, underlying relief on the

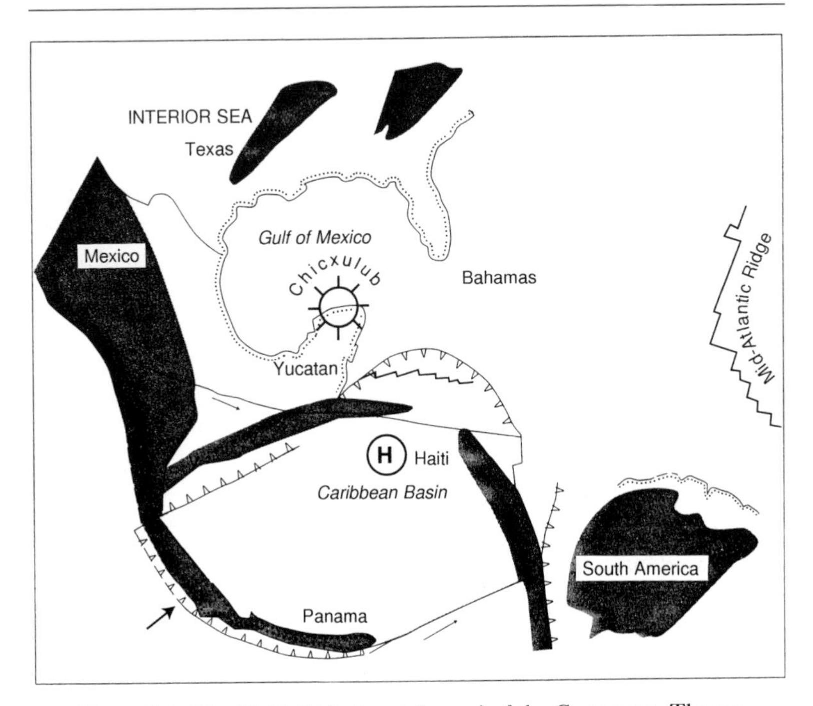

Figure 5.5. The Gulf of Mexico at the end of the Cretaceous. The sea level was higher than at present: continental seas flooded low-lying Yucatan and the North American plains. To the east of the Yucatan, a rift (jagged line) enlarges a small tectonic plate. To the south, Haiti (H) and its plate section glide eastward along a strike-slip fault. Central America at the time is a volcanic island arc bordering a deep-sea trench. On the Yucatan platform the emplacement of Chicxulub Crater is marked by a circle. (*Modified from Hildebrand* et al., *1991*.)

magnetic survey maps, with a hint of external, circular anomalies. Besides this promising feature, Hildebrand also noticed another possible crater candidate in northern Yucatan, although it was located slightly outside his search perimeter.

The feature in the Columbian Basin did not hold up to closer scrutiny and was soon rejected as a possible impact site. The Yucatan structure, on the other hand, turned out to be a completely different story.

The tale of the lost crater

The circular feature in the Yucatan had been known for years in the oil industry. In the early fifties, geophysical surveys had shown the buried structure to be 180 km in width, centered near the coastal town of Progreso. It extended half on shore, under the limestone plains of the Yucatan, and half offshore, under the water and sediment of the Gulf.

Circular features of this kind are of great interest to oil companies, because some circular structures can trap oil and gas. Surveys are conducted to identify those potential oil 'traps' worth exploring, like sedimentary anticlines, and to eliminate structures that are less likely to contain oil, like volcanic calderas.

With this objective in mind, geologists from the Mexican oil company Pemex drilled exploratory wells in the early fifties at half a dozen sites over the Yucatan feature. Hopes of finding oil were dashed when, after traversing hundreds of meters of limestone and other sediment, the drills struck a basement of crystalline rock at a depth of approximately 1500 meters, and the samples brought back to the surface were classified as volcanic andesite. To the disappointment of the explorers, the Yucatan structure appeared to be just another volcanic caldera.

A second set of exploratory wells, drilled in the late sixties, did little to change the volcanic verdict. But there was one geologist at the time, Robert Baltosser, who did make the suggestion that the structure could be an impact crater, based on the gravity data.

In 1978, a new geophysical campaign was conducted by Pemex across the Yucatan, including an airborne magnetic survey over the feature. American consultant Glen Penfield, who was gathering the data aboard the survey plane, was struck by the strong magnetic anomaly that defined the center of the structure – close to 60 kilometers in diameter – and by the outside ring of contrasting low magnetism that gave the feature its overall diameter of 180 kilometers. Could this pattern indeed be the mark of an impact crater, with the high magnetism corresponding to the central impact melt, and the contrasting low to the surrounding, poorly magnetized breccia?

Figure 5.6. In the late fifties and sixties, Pemex (Mexico's national oil company) drilled exploratory wells on the Yucatan platform to study a large circular feature buried one thousand meters below ground level. The feature was first thought to be a giant volcanic caldera. The sealed head of drill well C-1 still protrudes in the underbrush, near the village of Chicxulub. (*Photograph: by the author.*)

Figure 5.7. Impact breccia retrieved from the Yucatan-2 (Y-2) drill well, 500 meters below ground level. The Y-2 well is located outside the periphery of the crater, and samples the ejecta blanket on its outer slope. The breccia unit is 600 meters thick, and consists of fragments of gypsum, anhydrite and calcite, shattered and mixed together by the impact. (*Photograph: by Glen Penfield, Carson Services Inc.*)

There was also the gravity data. Gravity in northern Yucatan showed a rise of a few milligals[5] in the center of the structure, that could be interpreted as the influence of the dense, uplifted rock of the central peak, and an outer negative anomaly of twenty milligals, that could be caused by the ring of low density breccia.

Even though it had little relevance to the oil survey at hand, the discovery of a hidden impact crater – especially of such size – was big news, and Glen Penfield put the suggestion to Pemex in 1978,

[5] A milligal is a small unit of gravity that represents an acceleration of one hundredth of a millimeter per second squared.

two years before the announcement of the Alvarez *et al.* cosmic hypothesis.

With the permission by Pemex to release the proprietary information, Glen Penfield and Pemex geologist Antonio Camargo[6] made a more formal presentation of their findings at the yearly convention of the Society of Exploration Geophysics, in October of 1981 in Los Angeles. In their presentation, Penfield and Camargo went so far as to propose a link between the 180 km wide impact structure and the great mass extinction of the dinosaurs at the end of the Cretaceous.

As it happened, the news spread rapidly, partly thanks to science reporter Carlos Byars of the *Houston Chronicle*, who was present at the geophysics convention. Byars grasped the importance of the crater report, and the story ran on the front page of the *Houston Chronicle* on December 31, 1981, highlighting the possible link between the Yucatan crater and the death of the dinosaurs. The story was later picked up in the March 1982 issue of *Sky and Telescope*.

The proposition of a large impact crater in the Yucatan also circulated around the K-T community. Walter Alvarez himself heard of the Yucatan feature, and tried to check out its impact origin, but he was quickly discouraged by the lack of drill cores available for study. Based in Texas, Glen Penfield discussed the issue at length with impact specialists at the Johnson Space Center. Houston reporter Carlos Byars kept signaling the case to whoever would listen. One person who did listen, at a 1990 science meeting in Houston, was graduate student Alan Hildebrand.

Ground zero: Chicxulub

Fresh from his Haitian discoveries, Hildebrand needed little convincing to take a closer look at the Yucatan structure. He tracked down and read the report by Penfield and Camargo, and met with

[6] Antonio Camargo Zanoguera, geophysicist at Pemex, was born in northern Yucatan on the very site of the buried feature.

Glen Penfield. What convinced him even further was a paper on the geology of the Yucatan by Lopez Ramos, which he and Glen Penfield translated from the Spanish.

The next step was to find the cores drilled by Pemex in the fifties and sixties, and look for mineral and stratigraphic evidence of impact. Hildebrand was able to locate some of the cores at the University of New Orleans. The first cores he recovered, with the help of Penfield, were from the Yucatan-2 drill site, well outside the would-be astrobleme. Collected at depths between 1200 and 1300 meters, the samples were of shattered and mixed breccia rock, containing grains of shocked quarz up to one centimeter in size. This was evidently the ejecta blanket from the impact.

Next, Hildebrand recovered cores from inside the perimeter of the would-be crater, at the Yucatan-6 drillhole, supplied to him by Antonio Camargo. The samples displayed the characteristic mineralogy of impact products, including shocked mineral grains.

Hildebrand had no access to samples from other drill sites (the rumor was that their storage hangar in Mexico had burnt down in some mysterious fire), but written logs of these sites had fortunately been preserved.

For the central C-1 well, in particular, the logs described a melt sheet several hundred meters thick at the bottom of the drillhole, in keeping with the predicted thickness of impact melts in large astroblemes.

From the logs one could also construct a cross-section of the 180-km-wide crater by drawing curves between corresponding units at all the different wells. This showed that inside the rim of the crater, the first limestone beds to form in the underwater basin slope down one thousand meters from rim to bottom. This was the right profile expected of a 180-km-wide impact basin.[7]

Ten years after Penfield and Camargo, Alan Hildebrand thus realized that he had identified a giant impact crater, and a prime candidate for the K-T layer and extinction. But convincing the scientific establishment was quite another matter.

As a student, Alan Hildebrand met with the disbelief of more

[7] Computer simulations indicate that after reaching a maximum depth of over 15 km at the moment of impact, the bowl of such a large crater rapidly readjusts by central rebound, peripheral landslides and ejecta fall back, to reach an equilibrium depth of 1000 meters.

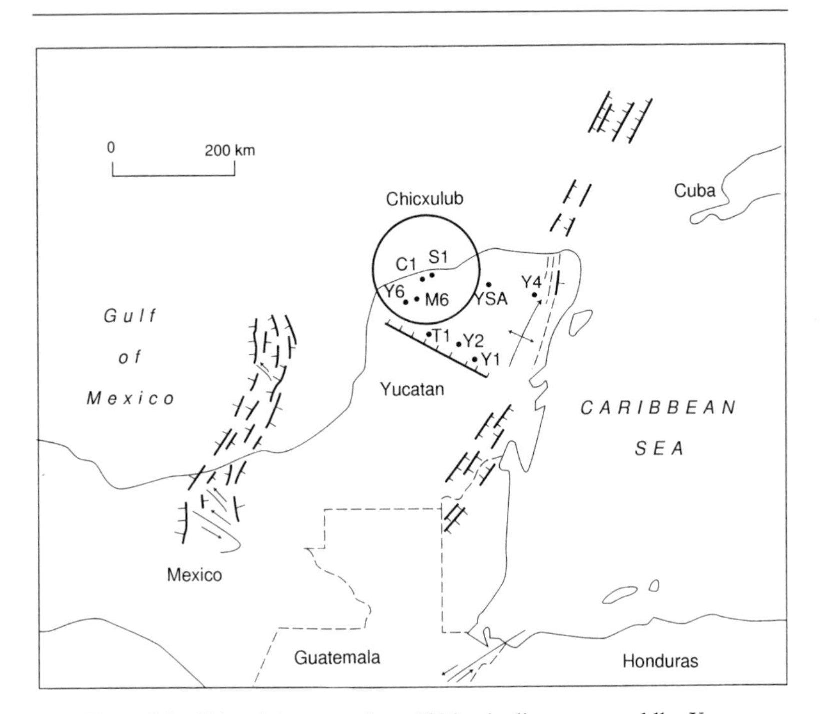

Figure 5.8. Chicxulub crater, about 180 km in diameter, straddles Yucatan's northern coastline. The impact basin is buried under a thousand meters of sediments. Its northern half stretches under the Gulf, and its southern half lies under the low-lying peninsula. Faults and folds are indicated by solid lines. The Pemex exploratory wells are indicated (C1: Chicxulub-1; S1: Sacapuc-1; Y1: Yucatan-1, etc.). (*Modified from K. O. Pope* et al., *1993.*)

senior scientists in the field, who had preopted for a continental U.S. impact site, and did not wish to think otherwise. They had set their sights on the 35-km Manson Crater in Iowa, and it would be another three years before Manson Crater would be redated at 73 million years, and discharged of the K-T massacre.

Alan Hildebrand published a brief note on the Yucatan crater in a 1990 issue of *EOS* science magazine, but met with much more resistance when it came to publishing a full-fledged report, co-written with his thesis director William Boynton, geophysicist Mark Pilkington, geochemist David Kring, Glen Penfield and Antonio Camargo.

They first submitted a paper to *Nature* in May of 1990, but this was before they found shocked quartz in the cores. Partly because of the absence of shocked quartz in their report, the paper was turned down. This calls for a little digression on the selection process of scientific articles.

Research magazines of the caliber of *Nature* select articles on the basis of peer review. Submitted articles are read by a panel of specialists in the field (usually three), and the decision for publication, rejection or revision is reached by a majority vote. Unfortunately for Hildebrand *et al.*, reviewers had little sympathy for the idea of a continental Mexican crater, which clashed with the prevalent ideas that the crater was oceanic, or else in the continental USA (Manson). The article submitted to *Nature* was rejected by two votes to one.

Soon thereafter, Hildebrand and his coauthors submitted their article to an excellent American journal of the Earth sciences, *Geology*. This time, they were able to report that nearly a third of the quartz and feldspar grains in the breccia rock from the crater were shocked. One influential reviewer was again hostile to a Mexican crater, and recommended against publication. However, both other members of the committee were convinced by the data and its interpretation, and the Hildebrand *et al.* article was published in the September 1991 issue of *Geology*. This is when the science community was formally advised that a giant impact crater had been discovered at the K-T boundary, and went by the unlikely, and nearly unpronounceable name of Chicxulub.

The name Chicxulub (pronounced Chic-shoe-lube) was chosen in reference to a small fishing harbor close to ground zero, Puerto Chicxulub.[8] Alan Hildebrand and his partners found the local name Chicxulub fitting for a crater on Mayan territory. There is no consensus yet for the meaning of the name *Chicxulub* in the Mayan language, although some say that it could mean 'sign of the horns' or 'tail of the devil', which would be very allegorical indeed.

[8] There is also a larger, inland village to the south, the Chicxulub Pueblo proper. It is next to this village that Pemex drilled the C-1 (Chicxulub-1) well in the early fifties. The C-1 well is the closest to ground zero (15 km).

The sacred wells

With the publication of the *EOS* and *Geology* articles, the ball began to roll. In a letter published in *Nature*, remote-sensing experts Kevin Pope, Adriana Ocampo and Charles Duller pointed out that one could see a hint of the crater's circumference on satellite imagery, in the form of a ring of small ponds.

Ponds are ubiquitous throughout the Yucatan. They are typically hundreds of meters in diameter, tens of meters deep, and are called *cenotes*, 'sinkholes' in Mayan (see figure 5.9).[9]

It so happens that around Merida, a great number of these *cenotes* are aligned in a broad semi-circle 80 kilometers in radius. The alignment is best seen from the air, especially to the southwest of Merida. In this narrow swath less than three kilometers wide, there is a record concentration of three ponds and sinkholes per square kilometer. One interpretation is that the ring of *cenotes* is caused by a major fault – the crater's boundary fault – that intercepts the flow of groundwater and causes it to upwell and undercut the surface.

Water flows subterraneously in the Yucatan. The permeable limestone soaks up the rain, and the water percolates underground to the Gulf through permeable strata, following a slight declivity to the north. Where it meets the ring fault that bounds the crater, the water is apparently diverted up and around the obstacle, dissolving the overlying strata and forming caves and sinkholes up to the surface.

The *cenote* ring does not come to an end at the coastline. It can be traced under the sea, where springs of fresh water well up from the sea floor and form noticeable thermal anomalies at the surface, visible on infrared imagery. In the coastal sounds, the local fishermen call these upwellings *'ojos de agua'* – aquatic eyes – in reference to the lenses of clear water that break to the surface.

[9] *Cenotes* were used as sources of drinking water by the Toltecs and Mayas. Some were worshiped as sacred wells, as was the *cenote* of the Mayan city of Chichen-Itza, which was the site of human sacrifices.

Figure 5.9. Yucatan's limestone platform is riddled with sinkholes, caused by the dissolving action of ground water. Known as 'cenotes', these ponds have served as drinking holes and bathing pools to the Mayans since times immemorial. Represented here is the 40 meter-deep Xlacah cenote, near the Mayan ruins of Dzibilchaltun. Around the city of Merida, the cenotes are aligned in a semi-circular belt that betrays the underground ring fault of the buried crater. (*Photograph: by the author.*)

The summer of '92

At the beginning of 1992, there was little doubt that the Chicxulub structure was an impact crater, and one of the largest ones on Earth, only matched in size by the much older Sudbury and Vredefort craters in Canada and South Africa. On the other hand, proving beyond a doubt that it was the source of the K-T boundary clay – and thus the culprit in the great mass extinction – was a more difficult matter.

Buried under a thousand meters of limestone, the structure

would not easily give up its secrets. Unless more exploratory drilling were performed – which was unlikely in the short term, given the cost of such surveys – K-T sleuths were left with a limited set of cores dating back to the fifties and sixties in order to prove their point, i.e. establish a chemical and temporal link between the Chicxulub crater and the great End-Cretaceous mass extinction.

In their *Geology* paper of 1991, Alan Hildebrand *et al.* had analyzed a sample from the Y-6 core, identifying it unambiguously as an impact melt. Other analyses followed, confirming the impact nature of the Chicxulub melt rock and breccia, and pointing out their chemical likeness to the microtektites found in the K-T sediments around the Gulf of Mexico.

Of course, one could always claim that the chemical match was only a coincidence, and that the K-T spherules and the Chicxulub crater were nevertheless two different events, separated in time and unrelated. On the other hand, a careful dating run on samples from both the crater and the K-T layer would most likely settle the issue: if the impact melt at Chicxulub and the K-T spherules at Haiti boasted exactly the same age, then the coincidence in time, added to the chemical match, would unambiguously convict Chicxulub crater as the source of the K-T layer.

The study was undertaken by a team of twelve scientists under the leadership of Carl Swisher, expert in argon–argon dating[10] at the Institute of Human Origins in Berkeley, and including K-T experts Alessandro Montanari (Italy), Jan Smit (Holland), Philippe Claeys (Belgium), Walter Alvarez (USA), and José Grajales-Nishimura and Esteban Cedillo-Pardo (Mexico).

The scientists sifted through the Chicxulub cores in their possession and picked three beads of glass from the C-1 drill well, sufficiently well preserved to yield reliable age estimates. After

[10] Dating rocks millions of years old is not an easy task. One technique consists in measuring the proportions of different atoms in their make-up, which are known to disintegrate at a fixed rate. From the estimated quantity of a radioactive element in the rock when it was formed, the quantity measured today, and the known rate of decay, one can calculate the fourth parameter of the equation: the duration of the decay process, i.e. the age of the rock. Although relatively simple in theory, this radiochronology technique is difficult to perform: atomic quantities measured are minute and subject to sampling and measurement errors. The most recent argon/argon technique limits the age error to less than half a percent, and sometimes down to a tenth of a percent – which corresponds to an age uncertainty of only 65 000 years for a 65 million-year-old rock.

performing the intricate argon gas measurements and processing the data, they obtained three independent ages, one for each bead of glass: 64.94 ±0.11, 64.97 ±0.07 and 65.00 ±0.08 million years. Averaging the results, the team pinned down the age of the Chicxulub crater to 64.98 ±0.05 million years.

They then dated the Haitian spherules and found 65.01 ±0.08 million years, perfectly matching the 64.98 ±0.05 age of the crater. Hence, the two events were not only chemically, but also temporally indistinguishable.[11]

Mixed reactions

These matching dates, published in the August 1992 issue of *Science*, sent an electroshock rippling through the scientific community. Chicxulub crater gained credibility. Even scientists who had initially belittled the feature, preferring to it other impact candidates like Manson crater in Iowa, were instantly converted. Virgil Sharpton in particular, of Houston's Lunar and Planetary Institute, began referring to the Yucatan crater as 'the smoking gun' of the K-T extinction.

There was also another, independent confirmation. Bruce Bohor of the USGS and Thomas Krogh and Sandra Kamo at the Royal Ontario Museum pored over shocked zircons in the K-T layer – minerals that came from the target rock on the site of the impact. These mineral ejecta were found in the K-T clay from as far away as Colorado and Saskatchewan.

The team performed radiochronology dating of the zircons, this time by measuring the disintegration of their uranium to lead (a technique well suited to zircons). They found in each zircon grain two superposed 'event dates', two dates that could be untangled with some numerical processing. One was the 65 million year date, that corresponded to the shocking and heating of the zircons during the impact, and the other was an older age of 545 million years,

[11] The same year, Glen Izett and his team also dated the glassy spherules of the K-T layer in Haiti. They obtained an age of 65.06 million years, with an error bar of only 0.3 % (180 000 years), matching the Swisher *et al.* results.

which reflected the age of formation of the original mineral – the age of the crustal basement on the target site.[12]

It so happens that the Yucatan crustal basement is known to be roughly of that age, in contrast to crustal basement under the Manson Crater in Iowa, for instance, which is much older. This 'Yucatan' basement age of the K-T zircons sealed the conviction of Chicxulub as the K-T impact crater.

Only a minority of scientists were left unconvinced. In the January 1994 issue of *Geology*, a rebuttal claiming that Chicxulub was volcanic was co-signed by John Lyons and Charles Officer of Dartmouth College, and Arthur Meyerhoff – consultant to the Pemex drilling campaign in the sixties. Meyerhoff reiterated his past interpretation of the Yucatan cores, and that the igneous rocks at Chicxulub were volcanic lavas, separated by layers of limestone marking long gaps between eruptions. The authors stressed that these 'volcanic episodes' displayed substantial chemical variability whereas, in their belief, impact melts were supposed to be rather homogeneous. As for the shocked minerals in the Chicxulub cores, the volcanists continued to attribute them to explosive eruptions.

The arguments put forth by the volcanists were not convincing. The so-called interstratification of 'volcanic and sedimentary' layers was probably nothing more than the mixing of pods of impact melt with slumped sediment at the edge of the crater. Neither did the chemical variability of the igneous rocks speak against an impact origin. On the contrary, impact specialists had long shown that in large astroblemes, melt lenses were chemically diverse, because of the variety of rocks that are fused and mixed together.[13] As for the shocked minerals, their deformation in criss-crossing planes was symptomatic of impact and not of volcanism, an interpretation which no longer suffered any contest among qualified specialists.

The volcanists' last stand was therefore dismissed, with no need for further debate. Some scientists were even surprised that a jour-

[12] There is a third age, 420 million years, that shows up in the mathematical treatment of the data, probably representing a different layer of crustal basement on the site. This variability is not surprising in view of the 180 km dimension of the crater.

[13] Melt rock at Sudbury crater in Ontario ranges from diorite to andesite and gabbro.

Figure 5.10. The identification of Chicxulub crater was met initially with reserved skepticism, until its impact mineralogy and K-T age were clearly established in 1993. Walter Alvarez (left) confers with the original discoverers of the crater: Glen Penfield (center) and Antonio Camargo (right). (*Photograph: by the author.*)

nal as serious as *Geology* accepted it to publish. In fact, the journal prudently listed the Meyerhoff *et al.* communication as an 'opinion', and not as a research article in the true meaning of the word.

The impact origin of Chicxulub crater became so compelling that at the end of a symposium held in Houston in March of 1994, which brought together over one hundred K-T specialists, including critics of the cosmic theory, guest lecturer Walter Alvarez pointed out that not a single person had taken the floor over the course of the meeting to contest the impact origin and the K-T age of the Chicxulub crater.

A new look at the cores

To impact specialists, Chicxulub crater was a bonanza. Not only was the K-T astrobleme finally discovered, not only did it turn out to be one of the largest craters on Earth, but it was also remarkably well preserved.

Most continental astroblemes are subject to erosion from the day they are formed, and lose their original features in a few million years. In contrast, because the impact took place on a submerged continental platform, Chicxulub crater was rapidly buried under a blanket of coastal sediment and shielded from erosion. The preservation was all the more effective in that the Yucatan platform has not undergone any tectonic deformation since K-T time.

The downside was that the crater was buried! Its protective limestone cap was a thousand meters thick. Besides the geophysical surveys and the Pemex cores, new data would be difficult to retrieve, and future exploration campaigns would also depend on the interest and good will of the Mexican government.

Working relations were set up between U.S., Canadian and Mexican Institutions. The Canadian Geological Survey established links, starting in 1991, with Mexican researchers at UNAM and UADY Universities, IMP and Pemex. In Houston, the Lunar and Planetary Institute also established relationships across the Gulf, and set up a joint program to explore the crater.

More confirmations and breakthroughs followed. In a 1993 article co-written with six other researchers, Joel Blum of Dartmouth did isotopic analyses of both melt rock from the crater (coming from the C-1 drillhole) and the Haitian K-T spherules. The analysis showed the isotopic compositions to be indistinguishable and confirmed the genetic link between the crater and the K-T ejecta.

In another paper dealing with shock metamorphism, Sharpton *et al.* confirmed the earlier results of Hildebrand *et al.* that a third of all minerals present in the crater breccia were shocked. They reported two distinct levels of deformation in the minerals, corresponding to shock pressures of 6–10 GPa and 20 GPa respectively.

One explanation was that these were separate ejecta components that originated from zones in the target area that were differently affected by the blast.

Tidal upheaval in Mexico

In parallel with the exploration of Chicxulub crater, research continued around the Gulf of Mexico to further identify and size up the ejecta blanket and tsunami deposits. In 1992, a study by Walter Alvarez, Jan Smit, Alan Hildebrand *et al.* focused on two underwater sites, 500 km from ground zero, between Yucatan and Florida. Deep sea drilling had recovered cores of sandstone two to three meters thick at the K-T boundary, bearing shocked quartz, spherule beds and a sprinkle of iridium, overlying a thick basement of gravel-rich clay. The investigators interpreted the sandstone to be ejecta-related, and the gravelly clay to result from underwater landslides triggered by the impact.

Just as exciting was the discovery by Jan Smit, Sandro Montanari and Walter Alvarez[14] of outstanding outcrops above sea level, in the Gulf provinces of northeastern Mexico. There, the uplifted Cretaceous strata showed a sandstone layer three meters thick at the K-T boundary, apparently running for hundreds of kilometers through the Tamaulipas and Nuevo Leon provinces. Where it was bared by erosion and accessible to geologists – notably on the sites of Mimbral, Mulatto and El Peñon – the boundary deposit consisted of a lower bed of spherules, half a meter thick and laden with shocked quartz, that was interpreted to be the early ejecta from the crater; a middle bench of thick sandstone, with abundant coastal debris (fish teeth, fossilized vegetation), thought to represent the coastal sediment reworked by the powerful tsunamis (see figures 6.2 and 6.3); and an upper unit of finer clay with an iridium-rich horizon, its ripple marks engraving in mud the oscillations of the abating currents (see figures 5.1 and 5.2).

[14] Walter Alvarez tells of the exciting discovery of one such site, Mimbral, in his book *T. rex and the Crater of Doom*, Princeton University Press, Princeton, 1997.

Besides bringing new evidence of a tsunami catastrophe along the Gulf coast, the Mexican sites helped to independently confirm the location of the source crater, by way of triangulation. Alan Hildebrand and John Stansberry plotted the location of the spherule beds of Mimbral and El Peñon, and taking into account the equally thick spherule beds of Texas and Haiti, narrowed down the source of the ejecta to northwestern Yucatan – indeed the site of the Chicxulub crater.

Structure of the crater

In their cosmic hypothesis of 1980, Alvarez *et al.* had calculated from the amount of iridium dispersed around the globe that the source crater measured 150 to 200 km in diameter. That the Chicxulub astrobleme apparently measured 175 to 180 km across was a stunning answer to their predictions.

In 1992, in order to better constrain the size and structure of the crater, Alan Hildebrand and geophysicist Mark Pilkington, with the help of Carlos Ortiz at the Mexico Institute of Geophysics, compiled the existing gravimetric and magnetic data (including new measurements of their own) with two seismic profiles across the basin collected by Pemex, and the stratigraphic information provided by the C-1, S-1 and Y-6 drill wells.

Most prominent was the gravimetric signature of the crater. It showed a semi-circular outline, 180 km in diameter, opening up toward the north-west in a horseshoe pattern (see figure 5.11). Gravity inside the crater was 30 milligals lower than the regional average, except for the very center of the structure where the gravity figures surged back up to a near absence of anomaly.

As we saw previously, this bull's eye pattern is typical of impact craters and betrays the abnormal density of their rock units. Low gravity, 'negative' anomalies are believed to arise from the hundreds of meters of shattered rock inside the impact basin: their density being lower than that of the surrounding, unbroken rock, one gets a lower reading for the local gravity field. As for the positive anomaly at the center of the structure, it is attributed to the uplift of dense

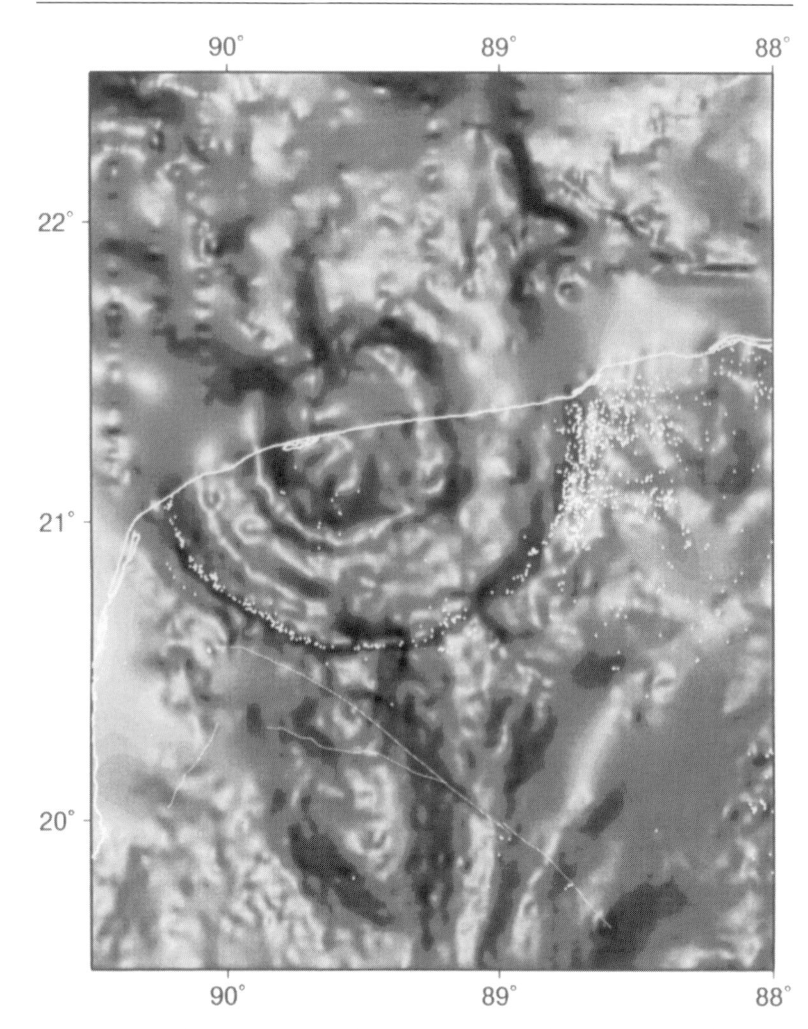

Figure 5.11. Map of the gravity anomalies across the northern coastline of Yucatan (white line). The data plotted is the horizontal gradient of the local gravity field. The central, dark positive anomaly is caused by the dense rocks uplifted by the impact, and the brighter ring of negative anomalies by the lower density breccia surrounding the peak and filling the basin. The periphery of the crater is highlighted by a string of cenote ponds (white dots), conspicuous in its southwestern quarter. The Y-shaped line in the lower third of the image is the Ticul fault. (*Photograph: courtesy of Alan R. Hildebrand, Natural Resources Canada.*)

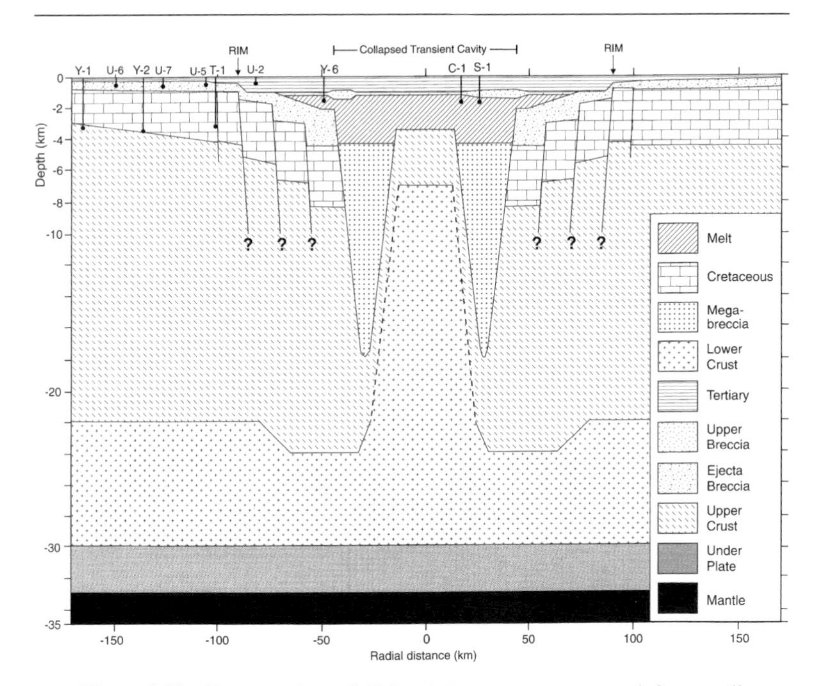

Figure 5.12. Cross-section of Chicxulub crater, constructed from well logs, cores and geophysical data. The vertical exaggeration is 8×. Exploratory wells by Pemex are drawn, as well as the inferred ring faults that bound the slump terraces (vertical lines with question marks). The model features a thick impact melt (hatched lines) covering the principal breccia units and the uplifted crustal rock. A ring of ejecta thins outwards, draping over the ring faults. (*Courtesy of Mark Pilkington and Alan R. Hildebrand, Natural Resources Canada.*)

basement rock close to the surface, that locally increases the gravity field.

The geometric center of the bull's eye pattern – and by inference of the crater itself – lies at 21.27 °N and 89.60 °W, near the small fishing village of Puerto Chicxulub. The southern half of the crater lies under the brush and agave plantations of the peninsula, while the northern half stretches under the shallow waters and sediments of the Gulf.

Besides the gravity data – which defines the overall geometry of the crater – the magnetic data further refines its structure by highlighting the boundaries of its buried impact melt. As it cools, molten rock indeed has the property of 'freezing' in place the magnetic field of the period.[15] At Chicxulub, this 'ghost' field is all the more detectable in that the surrounding sediment is very weakly magnetized, which makes for a sharp contrast. Thus the central anomaly is clearly visible as a ring-shaped feature lying 20 to 45 km from the center of the crater. Processing the data shows that the source of the anomaly lies most likely at a depth of 1100 meters, which is indeed the depth at which impact melt shows up in the drill cores. From this, Hildebrand and Pilkington deduced that the ring of impact melt had an outer diameter of 90 km, a dimension consistent with the size of melt pools observed in other astroblemes of similar size (such as Sudbury in Ontario). Based on these figures, the formation of Chicxulub crater could be described as follows.

At the moment of impact, the cosmic projectile blasted an initial cavity about 90 km in diameter and 30 km in depth. Lined with boiling impact melt, this transient cavity immediately began to collapse and grow outwards, great landslides widening and shallowing the crater to a final diameter of 180 km. As a rain of debris fell from the heavens onto the shifting floor, the heart of the structure surged upwards to form a rebounding central plateau. By the time the crater had reached a stable profile, minutes after the impact, landslides and ejecta fill had shallowed its depth to less than a kilometer.

In terms of volume of material created and displaced, Hildebrand and Pilkington estimate that the ring of impact melt in the middle of the crater is 2000 to 3000 meters thick, which makes for a total volume of 20 000 km^3. The central uplift, which underwent about 20 km of vertical rise, probably broke through the melt pool straight to the surface. As for the ejecta fallback that covers the melt, it also measures 3000 meters thick in the central part of the crater, and thins out to about 600 meters on the periphery – perhaps somewhat less in those places where the rim was eroded away by the action of the sea.

[15] In the End-Cretaceous, at the time of the impact, the magnetic field happened to be reversed, with the magnetic north at the south pole and vice versa.

A question of size

The 180-km model of Hildebrand and Pilkington satisfied all the data available at the time: gravity and magnetic surveys, seismic profiles, drill core stratigraphy, thickness of ejecta blanket and amount of iridium strewn around the globe.

In 1993, however, two other models were proposed for the Chicxulub crater, each suggesting a different size. Kevin Pope *et al.* favored a 240-km crater and Sharpton *et al.* a 280-km monster. This disagreement as to size was no trivial matter since the 'enlarged' models implied impact energies up to ten times those of the original model – a sizeable difference in terms of environmental disturbance and mass extinction potential.

In their 240-km model, Kevin Pope and his coauthors offered an alternative interpretation of the *cenotes* ring – the semi-circular alignment of sinkholes. They suggested that it represented not the outer edge of the crater rim, as implied by Hildebrand and Pilkington, but the inner boundary of the terraces leading up to the rim. If this rim extended a further 30 km outward, then the diameter of the astrobleme jumped from 180 km to 240 km.

Virgil Sharpton and his team had an even larger vision of the crater. By sifting through the gravity data, they saw evidence for additional anomaly rings, stretching farther out from the center of the crater. They interpreted the anomaly at the *cenotes* ring to be not the rim boundary, but the edge of the impact melt – doubling the diameter of this central unit from 90 to 180 km. In this model the rim of the crater was much farther out, at the level of a subdued anomaly ring boasting a diameter of 280 km.

The 'enlarged' models of Pope and Sharpton made a splash, both in the K-T arena and in the media, which was especially receptive to enlarged, more spectacular views of the catastrophe. But crater specialists remained skeptical. The magnetic data, after all, did confine the buried melt rock to a diameter of 90 km. Along with core stratigraphy, seismic profiles and calculated volumes of ejecta, the whole data converged to support Hildebrand and Pilkington's model, rather than the larger ones.

Hildebrand and Pilkington themselves, as well as independent specialists like Richard Pike, reviewed the gravity analysis of Sharpton *et al.* and did not find any convincing trace of an outer anomaly ring at a diameter of 280 km. But the best way to settle the issue was still to return to the Yucatan and collect new data.

Back in the field

By early 1994, Chicxulub had become the target of several exploration programs, each tackling a different issue and supporting a different model for the crater.

Remote-sensing experts Kevin Pope and Adriana Ocampo pursued their analysis of the *cenotes* ring and of ground water flow, as did Yucatan specialists Eugene Perry and Luis Marin.

Building up the international cooperation between Houston and Mexico City, Virgil Sharpton and Luis Marin set up a program of shallow drilling on the edge of the crater, in order to recover new cores. The light-weight drilling rigs only allowed penetration to depths of 700 meters. The deep center of the basin was therefore off limits and the wells were drilled mostly on the periphery of the crater where the interesting strata – melt pods and breccia – came sufficiently close to the surface to be sampled.

Drilling began in March of 1994, with the hope of finding evidence that would support the 280-km model. The interpretation of the shallow cores was that the buried relief of the crater's ejecta blanket climbed to a high stand – interpreted to be the crater rim – some 125 to 150 km from the center of the structure. Further analysis of the data convinced the Sharpton team that the Chicxulub 'multiring basin' was indeed 300 km in diameter, a figure that was later updated to a whopping 400 km. But to critics, there was really no tangible evidence for such a large basin, even in the newly-drilled cores.

While Sharpton's model kept growing, Hildebrand's vision of the crater remained stable at 180 km, as verified by a new gravity survey conducted by his team in the fall of 1994. Hildebrand and coauthors Pilkington, Connors, Ortiz-Aleman and Chavez enriched the existing data base with gravity readings that they collected along

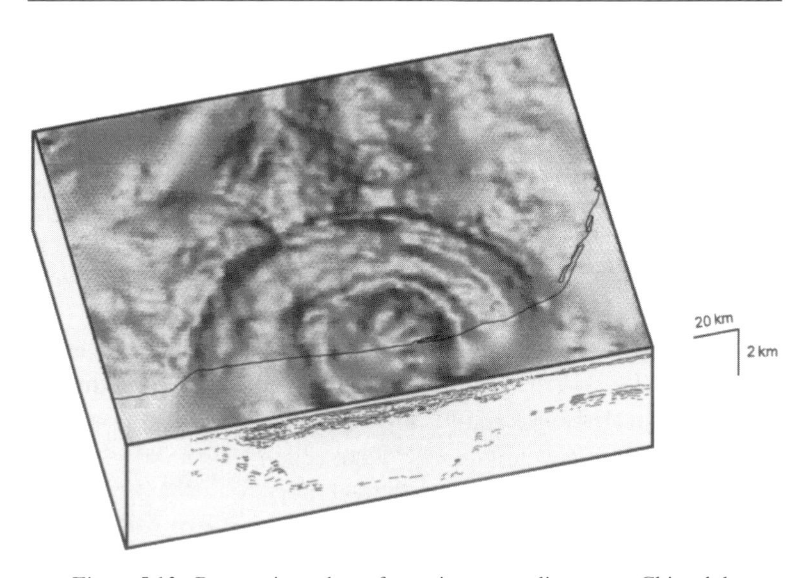

Figure 5.13. Perspective plot of gravity anomalies over Chicxulub Crater, combined with seismic reflection data by Camargo and Suarez (1994). Slump faults are noticeable inside the outer ring in the form of a wavy line, most prominent to the east (left). Vertical exaggeration is 10 ×, and the view is looking south. (*Photograph: courtesy of Alan R. Hildebrand, Natural Resources Canada.*)

five traverses radial to the crater. Each of the five traverses – four inland and one along the coast – was 40 km long, the surveyors stopping every few hundred meters to take a new reading. All in all, several hundred new data points were added to the digital gravity map, which was specially processed to highlight those features indigenous to the crater.[16]

In the new map thus obtained, Hildebrand *et al.* confirmed the presence of six anomaly rings between 20 and 90 kilometers from the crater center and none beyond, confirming their original 180-km model and refuting the enlarged models of Pope *et al.* (240 km) and Sharpton *et al.* (280–400 km). In addition, their own model gained substantial detail. They interpreted the two inner rings to

[16] Rather than plotting the gravity anomalies as such, Hildebrand *et al.* plotted their first derivative (i.e. the horizontal gradient – or change – of the anomaly over distance) which in effect does away with the regional 'background noise' of the gravity field and highlights the local trends directly associated with the crater.

mark the boundaries of the central uplift and of the surrounding moat of impact melt, while the next three next rings – 55 to 80 km from the center – marked the boundaries of slump terraces in the periphery of the basin, stepping up to the final sixth ring at 90 km radius, which represented the outer crater rim.

Moreover, this last 'rim' feature displayed an irregular outline, as revealed by slight variations in its radius from one traverse to the next. This scalloped appearance was typical of outer rims of large impact craters on the Moon, Mars and Venus.

The geophysical report by Hildebrand *et al.* was published in the August 7, 1995 issue of *Nature* magazine, which also featured the new Chicxulub gravity map on its front cover (figure 5.11) – a well-deserved tribute to the Canadian scholar and his team who had identified and modeled the great impact crater of the End-Cretaceous.

An international consensus

Today the exploration of Chicxulub crater continues unabated. Alliances are sought, and research plans are drawn to collect new data.

Such an international survey was conducted in the fall of 1996, and involved Imperial College, London, under the leadership of Mike Warner and Jo Morgan, the British Institutions Reflection Profiling Syndicate (University of Cambridge), the Institute of Geophysics at the University of Texas, under the leadership of Dick Buttler (with a grant from the National Science Foundation), and Alan Hildebrand and Mark Pilkington of the Geological Survey of Canada. Other participants included Leicester University, Mexico's UNAM University, and Houston's Lunar and Planetary Institute.

The aim of the survey was to constrain the size and structure of the buried crater. A research vessel sailed over the offshore portion of the crater, off the coast of Yucatan, to collect seismic, magnetic and gravity data. Simultaneously, a second research team on shore collected more data, and listened in with their wide-angle seismometers to the seismic blasting at sea.

Operated by the University of Texas, the Research Vessel Longhorn reached the crater site in late September of 1996, and deployed 35 automatic seismometers that sank to the Gulf's bottom, two thirds of them along a 310-km east–west line – roughly parallel to the Yucatan coast – and the rest along a nearly perpendicular line, northwest and out to sea.

Once the seismometers were in place on the sea bottom, an accompanying ship, the GECO Sigma, fired powerful blasts with underwater airguns, sending shock waves through the water column, straight into the sea floor. The waves that reflected off the various bedding planes and structural boundaries of the crater were recorded by the seismometers on the sea floor, as well as by the team on the Yucatan coast.

The seismic cruise was a success. The GECO Sigma and the R/V Longhorn shot and recorded over 2000 airblasts – along with their instructive echoes – as they sailed over the offshore part of the crater.[17] Combined with the data collected on shore, the picture that emerges is that of an asymmetric crater (asymmetry along the east–west axis), with the deep crust disrupted across a diameter of roughly 100 km. This would be the deep bowl initially formed by the impact (the collapsed transient cavity). Slumped blocks and a broad terrace are also seen on the seismic profiles, extending the structure to an outer diameter of 185 to 195 km.[18] Thus the original model of Alan Hildebrand and Mark Pilkington (180 km) is fundamentally correct. It sets the impact energy at 5×10^{23} joules, close to the figure advanced by Alvarez *et al.* in their original hypothesis. This is of fundamental importance to correctly ascertain the damage caused to the environment, and to model the lethal mechanisms that led to the great mass extinction.

[17] Of the 33 ocean probes, only one was lost at sea in the rough waters whipped up by Hurricane Josephine. There was also the loss of one tow-along magnetometer (was it nipped off by a shark?). A total of 99 seismic stations were deployed on land.

[18] This outer ring is more than just a slumped terrace and boasts a deep fault running down through the crust and into the mantle. Other faults extend outside the crater proper, including a possible ring feature 240 km in diameter that might have originated as a thrust fault.

6

Scenario of a catastrophe

Today the Chicxulub crater is universally recognized as the source of the K-T boundary clay – and as the likely cause of the great mass extinction that the clay represents. The challenge is now to ascertain which extinction mechanisms were at play during the impact, and why so many species were extinguished, while others survived in large enough numbers to eventually repopulate the Earth.

When they published their cosmic hypothesis in 1980, Alvarez *et al.* suggested that the most lethal effect of a large impact would have been the expansion of a giant cloud of dust in the atmosphere, blocking off the Sun's rays for months, and inhibiting photosynthesis of terrestrial and marine plant life. The eradication of plants and plankton on a global scale would then have led to the collapse of the food chains on land and in the oceans. This first model was followed by a variety of other killing mechanisms and doomsday scenarios.

One hundred million megatons

To better assess the causes of extinction, let us review the expected sequence of events as the bolide from outer space plunged through the atmosphere and struck the Earth.

The global intensity of the blast was equal to the bolide's kinetic energy, which is defined to be half the mass of the projectile multiplied by the square of its speed ($E = 1/2mv^2$). From the amount of iridium dispersed around the globe, we know the mass of the projectile to be roughly one thousand billion tons. As for its velocity, we know from celestial mechanics that most Earth-crossing asteroids and comets travel at speeds in excess of 20 kilometers per second, relative to our planet.[1] Plugging both figures into our equation yields an impact energy on the order of 10^{24} joules.[2]

This figure is staggering: it is equivalent to the global heat flow radiated by our entire planet in a thousand years, or else – a more gruesome comparison – to one hundred million megatons of TNT, *10 000 times the explosive energy of the world's entire nuclear arsenal*, American and Russian bombs combined.

Scientists modeling this tremendous blast also have some idea as to how the energy was partitioned between the atmosphere, the ocean and the solid Earth. For starters, a projectile ten kilometers in diameter traveling at 20 kilometers per second has so much momentum that it crosses the Earth's atmosphere in a split second without noticeably slowing down (in contrast to small meteorites that are considerably decelerated by the friction of air). In fact, calculations show that the atmosphere was literally blown away during the strike,[3] the displaced mass of air taking some 10^{21} joules of energy away with it – a mere one tenth of a percent of the bolide's energy. As for the few dozen meters of water on the site, they were likewise volatilized in less than one hundredth of a second, before the impactor rammed into the ground with the energy of one hundred million atomic bombs.

Hurricanes and tidal waves

Before we turn to the pulverization of the Earth's crust, let us first follow the shock waves that spread through the atmosphere and the Gulf waters.

[1] In effect, the relative speed of the K-T bolide could have ranged anywhere from 11 to 72 km/s, the lower value corresponding to an asteroid on an orbit closely matching the Earth's and being slowly caught up and 'absorbed' by our planet's gravitational field, while the higher figure is that of a head-on collision with a comet on a very eccentric orbit. For discussion's sake, most scholars choose a 'conservative' collision speed of 20 km/s, typical of most Earth-crossing asteroids.

[2] This order of magnitude was confirmed by the 1996 international survey of the crater (see end of previous chapter) which fixed the impact energy at 5×10^{23} joules.

[3] This leaves us with a problem concerning the ablation spinels. Were these formed from the skin of the 10-km object during the few tenths of a second of entry, when it blasted away the atmosphere around it, or else did the spinels form from fragments of the impactor that blew out of the crater like shrapnel after the impact, and retraversed the atmosphere?

Figure 6.1. Artist's view of the Chicxulub impact in the Gulf of Mexico. Only one tenth of one percent (0.1 %) of the impact energy was transmitted to the ocean and atmosphere, but that was enough to drive hurricane-force winds and giant tsunami across the Gulf. The overwhelming majority of the impact energy (99.9 %) was transmitted to the ground, and drove the dispersion of the ejecta and fireball. For a view of the resulting astrobleme, see figure 4.1 (*Photograph: courtesy of William K. Hartmann (© William K. Hartmann).*)

The concept of shock waves was first explored in 1980 by Emiliani, Kraus and Shoemaker, only a few months after the publication of the Alvarez *et al.* cosmic theory. From the laws of conservation of momentum, it was clear that the atmospheric disturbance began its outward rush at a velocity close to that of the incoming projectile, i.e. at a speed on the order of 20 kilometers per second (70 000 km/h). The momentous blast saw its speed dramatically decrease in a matter of minutes, as the ring of air grew in diameter and encountered an increasingly large atmospheric mass, braking its progress. In about ten minutes' time, wind velocities had fallen below 1000 km/h (the speed of sound), roughly 500 km away from ground zero. After about an hour, the blast ring had reached a radius of 1000 km, with wind velocities still stronger than the worst hurricane winds on record.

The wind speed kept falling with distance, but the eolian disturbance was far from over. The outward blast had created a vacuum in its wake, centered on ground zero, and it was not long before the winds reversed directions, the air flow streaming back into the giant depression. This 'two-way' hurricane effect is observed on a smaller scale during nuclear tests.

From these rough estimates, one can speculate that forests were flattened – and dinosaurs blown away – up to 1000 km from ground zero. Hence, the eolian catastrophe was localized, especially since ground zero was at sea. Hurricane-strong winds were probably limited to the Gulf Coast and Central America.

The destructive power of the tidal wave was also limited because the impact took place in less than 100 meters of water. Since the amplitude of a tidal wave cannot be greater than the sea depth at the point of disturbance, the swell probably started out no higher than 100 meters.[4] The disturbance also lost a fair amount of energy as it spread across the sea, but it is likely that the waves were still tens of meters tall when they reached the shores of North America,[5] and a few meters when they hit the coasts of Europe and Africa, on the far side of the Atlantic. Moreover, since these open waves grow taller when they reach coastal waters – surging 10 to 20 times their initial height – tremendous swells must have swamped the coastlines, flooding low lying plains and racing many kilometers inland.

In North America, the waves certainly raced up the shallow interior seaway, deep into the continent. Since the coastal plains of this inner sea were teeming with life – and a breeding heaven for many species – one can well imagine that countless dinosaurs and other animals were drowned by the rushing waters.

[4] This is very different from a large impact in the deep ocean, where the initial wave would be thousands of meters high.

[5] When they studied the Brazos River tsunami deposits in Texas, Joanne Bourgeois and her team estimated that the wave amplitude was 50 to 100 meters. Several factors might explain these high figures, such as the run-up of waves close to shore, or secondary impacts of bolide fragments and ejecta, skipping out into the deeper waters of the Gulf and raising additional waves.

The big shake-up

The oceanic and atmospheric upheaval – tsunamis and hurricane-force winds – added up to perhaps one percent of the energy unleashed by the impact. The remaining 99 % was converted into the melting and vaporization of the bolide and target, the dynamic ejection of the ejecta out of the crater, and the seismic shake-up that rang the planet like a bell.

The seismic jolt was extremely violent. The ground waves that rippled out from ground zero would have scored an awesome 10 on the Richter scale, dissipating one thousand times more energy than the strongest earthquakes on record. Besides triggering fault movement and landslides in the immediate vicinity of the crater, the shock waves certainly extended their influence far and wide across the globe.

M.B. Boslough and fellow scientists at the Sandia National Laboratory have shown through computer simulations that the area antipodal to the Yucatan – the southwestern Pacific – was notably affected, because the internal layering of the Earth acted as a giant lens to refocus the seismic energy there. Converging on this 'ground zero prime', the shock waves from the impact probably caused the Indian Ocean floor to heave up in swells reaching a full ten meters in amplitude, an awesome figure compared to the one-meter ground swells of the San Francisco 1906 earthquake!

Could such refocusing of energy at the antipode have melted the upper mantle there, setting off volcanic eruptions?[6] The Deccan Traps of India come to mind, even though they were thousands of kilometers away from the impact's antipode at K-T time. Some scientists still wonder if the impact might have started the Indian eruptions, or at least have amplified them.

There is no evidence, however, of such a relationship. For one, computer models show that the impact energy that converged at the antipode would have been insufficient, in theory, to generate any significant amount of magma in the crust or upper mantle. We

[6] Such a process has been proposed by geologist Ron Greeley to explain an interesting alignment on Mars, where Alba volcano in the northern hemisphere is antipodal to the giant Hellas impact basin in the south.

Figure 6.2. Geologists are now conscious that impacts at sea can be identified on the basis of tsunami deposits conserved in coastal rock. Here at Mimbral, 200 km inland from the Mexican shore, a tsunami unit is exposed at the K-T boundary: one meter of coarse debris, level with the geologist's hand, and two meters of finer, smooth sandstone above her head. The unit is capped by a layer of iridium-laden clay (see figure 2.7). (*Photograph: by the author.*)

also know today that the Deccan Traps began erupting two to three million years *before* the Chicxulub impact, so that the latter could not possibly have caused the former.

It appears that the impact did not even accentuate the ongoing eruptions. Where it can be found in the Deccan Traps, the K-T layer indeed sits in the middle of a ten meter thick sedimentary unit, suggesting that the impact coincided with a lull in the eruptive history of the area, rather than with an episode of renewed activity.

Figure 6.3. Close-up of the K-T tsunami deposit at Mimbral, Mexico. The numerous fossils of plant cellulose attest to the vast amount of coastal debris that was carried to a depth of 100 meters in the aftermath of the impact. Ripple marks can also be seen at the top of the deposit (see figures 5.1 and 5.2). (*Photograph: by the author.*)

The ejecta plume

Much greater than the global shake-up was the tremendous energy released locally by the excavation of the crater. Upon impact, the wave of compression that spread through the crustal basement of the Yucatan was followed by its decompression tail that sent the pulverized rock flying out in all directions from the expanding crater. It was this ejecta that redistributed most of the energy of the impact worldwide.

The blast of incandescent ejecta started out as a narrow cone 'firing back' along the bolide's incoming trajectory, then fanning out into a hemispheric bubble resembling the blast of a traditional explosion. The violent expansion of vaporized sea water further accelerated the plume of debris flying out of the crater.

The initial ejecta jetting out from the central funnel was rich in volatilized bolide material – a silicate and metallic vapor reaching temperatures of thousands of degrees. This central jet, representing roughly ten times the mass of the impactor, was the source of the iridium-rich 'fireball' that spread high above the atmosphere and wrapped around the globe.

On the wings of this central plume of molten and vaporized material, a spray of solid, pulverized ejecta also moved outwards, representing up to 300 times the mass of the impactor. Nearly 200 000 cubic kilometers of target rock were blown out of the widening crater in less than a minute – the gaping hole reaching a depth of 30 km before the excavation ended and the steep walls collapsed to form an enlarged, 180-km-wide saucer.[7]

Thousands of billions of tons of solid, molten and vaporized rock were thus propelled skyward during the minute-long blast, at velocities which ranged from a few hundred meters per second at the edge of the basin to several kilometers per second in the center of the fireball, above ground zero.

The great fire

The central ejecta plume was not only propelled at the highest speeds, it also benefited from a near vacuum overhead – the hole punched in the atmosphere seconds earlier by the bolide's entry – so that its upward flight was not slowed down by air friction. Better yet, since it was a cloud of molten and vaporized matter, its thermal energy fueled its own expansion, the billowing plume spreading outwards in all directions over the Earth's atmosphere.

[7] Most of the proximal ejecta from the crater is buried under more recent sediment, and can only be sampled by deep drilling. However, Adriana Ocampo, Kevin Pope and Alfred Fischer discovered a remarkable outcrop of ejecta uplifted along a fault and exposed in a quarry at Belize, 364 km from the center of the crater. There, a 15-meter thick deposit of rounded fragments of dolomite (magnesium limestone) attests to an ejecta layer, probably emplaced as a ground surge of rolling rock and gas.

Hot solid particles condensed out of this cloud in the vacuum of space, and in contrast to the airless launch, met with air resistance as they fell back towards the Earth. Their fiery reentry heated up the stratosphere all around the globe, the hot projectiles radiating away both their kinetic energy and what was left of their internal heat.

In 1990, cratering expert Jay Melosh of the University of Arizona provided a first estimate of the resulting heat pulse. Assuming that the rain of ejecta averaged 10 kilograms of material per square meter all over the Earth, and assigning to the projectiles an average reentry velocity of 5 km/s, Jay Melosh and his partners calculated a heating pulse on the order of 100 million joules per square meter of atmospheric section. Most of this heat would have dissipated at altitudes of 60 to 70 km, where the fine ejecta underwent the largest amount of aerodynamic friction and braking, as is observed to be the case with shooting stars and small meteorites.

In this doomsday scenario, the upper layers of the atmosphere lit up in a blazing inferno, discharging in a matter of minutes a radiation output on the order of 50 kilowatts per square meter – over 30 times the energy the Earth normally receives from the Sun. Only a fraction of this thermal outburst would have reached the Earth's surface, however, since half the energy was radiated upward (back into space) by the heated stratosphere, and part of the downward half was absorbed by the gases of the lower atmosphere on its way to the ground.

According to Jay Melosh *et al.*, as much as 10 kW/m^2 still made it to the surface: in an hour's time this would have raised the temperature of the soil to 400 °C – the heat level of an oven set on 'broil'. This sustained radiation would have ignited fires over the entire planet, fires which would have further increased the ambient heat. In this inferno, animals and plants were grilled to a crisp, and only those areas spared by lucky circumstances would have claimed any survivors. In fact, in the light of such a tremendous heat pulse, one might even wonder why there were any survivors at all.

One possibility, as Jay Melosh and his coauthors point out, is that any cloud bank overhead would have acted as a protective buffer. As they were vaporized into steam, the water droplets in the clouds would have absorbed a large fraction of the incoming radiation, so that a thick enough cloud cover could have locally neutralized the heat pulse, or at least considerably cut down its lethal effect. Surprisingly then, the weather patterns around the globe might have played a crucial role at the time of the impact. Those regions enjoying blue skies or starry nights would have been the most severely affected, whereas overcast areas would have been relatively spared.

The heat pulse model by Melosh *et al.* is but a first order approximation. The real story was certainly more complex, as pointed out by Alan Hildebrand, who prefers to the uniform spread of ejecta a scaled distribution in which the amount of debris – and thus the heat pulse – falls off exponentially away from ground zero. Hildebrand estimates that the Melosh average – 50 kilowatts per square meter – reigned at distances of approximately 5000 km from the point of impact. Farther out, the heat and temperatures decreased. Closer in, on the other hand, the heat flux built up to values probably 100 times greater than average within a 2500 km radius around ground zero, and 1000 times larger within 1000 km of the crater. The radiation that singed the Americas was therefore more on the order of *1 to 10 megawatts* per square meter – igniting the forests with great ease.

There is yet another level of complexity that affects the celestial bombardment: the asymmetry of the fallback curtain of debris. As Walter Alvarez remarked in a 1995 paper, as the stream of debris rose high above the Earth, the planet's rotation brought the areas west of ground zero under the brunt of the fallback, namely the Pacific Ocean and Asia, whereas regions to the east, like the Atlantic Ocean and Europe, were relatively spared – a model that appears to be borne out by the scarcity of shocked quartz and tektites in the European K-T clay.

An oblique impact

Besides a rotational bias, an even larger asymmetry in the ejecta distribution is expected from the trajectory of the bolide itself. Indeed there is no reason to expect that the incoming asteroid or comet struck the Earth at a vertical angle. The collision could have occurred at a grazing or oblique angle, with a tongue of fire and ejecta spreading across the globe in one predominant direction. In fact, the entire gamut of incident angles is equally probable, from vertical to tangential, the median expected value being 45°.

Before Chicxulub crater was even discovered, planetary geologist Peter Schultz of Brown University had stressed how an oblique impact might account for the apparent severity of the End-Cretaceous mass extinction. By simulating impacts at the NASA/Ames gun range – with hypervelocity shots fired into a variety of targets – Peter Schultz established that oblique hits transmitted a large fraction of their energy downrange of the point of impact, with fragments of the projectile bouncing and skipping amidst a tongue of ejecta and vaporized material (see figure 6.4). The fragmented bolide might then dissipate more of its energy in the Earth's atmosphere and worsen the environmental disturbance.

Although an oblique impact has the highest probability of occurrence at the K-T boundary, and offers a good explanation for the intensity of the damage to the biosphere, such an occurrence remains to be proven in the field. Fortunately, planetary geologists know what evidence to look for. Experiments in the lab. and crater fields on the Moon, Mars and Venus show that oblique impacts produce distinct shapes and patterns. Craters take on a slightly elongated appearance at impact angles lower than 45°. Moreover, while the uprange crater wall remains steep and well defined, the downrange wall tends to be somewhat downdropped, often breaking open in a horseshoe pattern. Central peaks of oblique impacts also tend to be located slightly uprange of the geometric center of the crater.

When Chicxulub crater was discovered and mapped, Peter Schultz was quick to point out that its gravimetric imprint is asymmetric. The pattern is pronounced and semi-circular in the southeast, and fades into a more open horseshoe pattern to the northwest (see figure 6.5). From this, Schultz concludes that the bolide flew in obliquely from the southeast, over the South Atlantic and South America, before hitting the Yucatan at roughly a 30° angle. The impact would have sent an elongated fireball and a salvo of secondary projectiles hurtling downrange across the Gulf of Mexico, straight into North America.

Such a model has the merit of explaining why the largest tektites

Figure 6.4. High-power gun experiment, simulating an oblique impact. The projectile, fired at a speed of 5 km/s, struck the metallic target at a 15 degree angle. Such an oblique impact could explain the morphology of Chicxulub crater (see figure 6.5). (*Photograph: by Peter H. Schultz, Brown University and NASA Ames Vertical Gun Range.*)

and grains of shocked quartz are found in North America, and why plant and animal extinctions appear to be particularly severe on that continent, due to tidal waves spreading in the interior sea from secondary impacts across the Gulf of Mexico, as well as more ejecta bombardment and firestorms.[8]

[8] Alan Hildebrand and co-workers also see evidence of an oblique impact, but along a different direction. From the offset between the center of the crater and the center of the central uplift, as well as from evidence of maximum compressional shearing to the northeast side of the crater, Hildebrand *et al.* suggest that the bolide flew in from southwest to northeast with an elevation angle of about 60 °. In this case, maximum damage would have been directed over the Caribbean.

Figure 6.5. The gravity anomaly pattern of Chicxulub crater is asymmetric. This has led Peter H. Schultz and others to suggest an oblique impact directed southeast to northwest. (*Photograph: courtesy of Alan Hildebrand and Mark Pilkington, Natural Resources Canada.*)

Global wildfires

Whatever the angle of impact, the fireball and the dramatic heat pulse caused by the reentering ejecta certainly singed the greater part of the biosphere. This is the picture at least in theory. But is there any evidence of a heat pulse in the field, branded in the K–T clay?

In fact, the evidence had been stumbled upon as early as 1985, before the theory of a heat pulse was even proposed, by chemists Wendy Wolbach, Roy Lewis and Ed Anders at the University of Chicago. They were analyzing the K–T clay at the time to identify

Figure 6.6. The study of carbon-rich particles in the K-T clay led Wendy Wolbach and others to suggest global wildfires at the K-T boundary. The amount of soot is equivalent to the burning down of at least half the forests and worldwide plant and animal life in the aftermath of the impact. (*Photograph: courtesy of Wendy S. Wolbach, Illinois Wesleyan University*.)

any noble gases from the bolide that might have survived the fiery impact. From their experience with other rocks, the chemists knew that such gases, if any were left at all, would have been trapped in the carbonaceous fraction of the clay. Hence, they turned their attention to the specks of carbon in the K-T samples.

As it was, Wendy Wolbach and her partners did not find any trapped cosmic gases in the K-T clay, but they did find unusual quantities of elemental carbon: over four times the amount found in the surrounding limestone, or – after scaling the quantities to take into account differences in sedimentation rate – several *hundred times* the normal carbon flux of the period.[9]

[9] The centimer-thick K-T clay is assumed to have settled and formed in less than a year, whereas the limestones above and below formed at much lower rates of a centimeter per 100 years. Hence, not only was there four times more carbon in the K-T layer, but its deposition being 100 times more rapid, there was a 400-fold acceleration in the supply – or 'flux' – of carbon at K-T time.

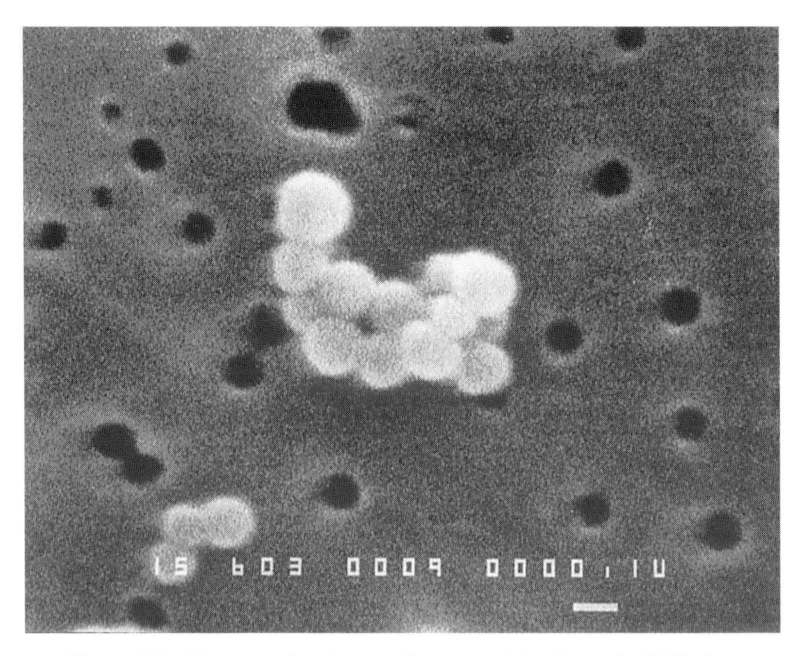

Figure 6.7. Close-up of a cluster of soot particles from the K-T clay (Caravaca, Spain). The carbon-rich globules are a tenth of a micron in diameter. Their arrangement in clusters is symptomatic of the combustion of organic material. (*Photograph: courtesy of Wendy S. Wolbach, Illinois Wesleyan University*.)

Not only was there a blatant excess of carbon in the K-T clay, but it also had an unusual appearance. Most specks were less than a micron wide, irregular and fluffy looking; some were spheroid and linked together in clusters (see figure 6.7). In fact, they looked just like the type of carbon generated by flaming combustion – soot or 'carbon black'. Chain-like clusters, in particular, were only known to form in flames, or very hot gases. There were also some larger carbon particles of irregular shape in the K-T clay, that clearly resembled charcoal from forest fires.

The carbon record of the K-T layer thus pointed to catastrophic fires ignited worldwide by the impact and the ensuing heat pulse. Before jumping to conclusions, however, Wolbach and her coauthors did sift through alternative explanations, but chemical considerations allowed them to rule out the possibility that the carbon might have been meteoritic, as well as the scenario that vast reserves of fossil fuel – like oil shale – might have been ignited on the site of the impact. The source of the K-T carbon had to be the biosphere itself: forests worldwide – and the animals they harbored – had gone up in flames.

In a follow-up study published in 1988, Wendy Wolbach and coauthors confirmed the earlier findings, and showed the carbon enrichment to be even larger on the site of Woodside Creek in New Zealand, where the carbon flux increased more than 3000-fold in the first few millimeters of the K-T clay.

Worldwide, the chemists estimated the amount of combustion carbon in the K-T layer to be on the order of 100 billion tons. Compared to the mass of vegetation assumed to have covered the Earth at the time (a figure actually larger than today's, since the Cretaceous was a period of luxuriant forests), Wolbach *et al.* concluded that over half of the world's biomass was burnt down in the aftermath of the impact, within a timeframe of days to months since most of the soot is concentrated in the bottom three millimeters of the K-T clay.

A confirmation of the combustive nature of the K-T carbon later came from the discovery of polycyclic aromatic hydrocarbons (PAHs) at the surface of the soot particles. These are heavy molecules that can be produced by the burning of organic matter. In the K-T clay one finds phenanthrene and coronene, pyrexene, fluoranthene and retene – the latter probably resulting from the thermal decomposition of pine tree resin.[10]

It should also be noted that on many K-T sites around the world, there is an abrupt change in the vegetation immediately after the impact. Whereas the forests before the crisis encompassed a variety of species – from conifers to flowering trees – the fossilized vegetation immediately above the K-T boundary consists essentially of ferns (up to 90 % of the total spore count). Ferns are resistant species that are the first to cover the land after forest fires. This is not unambiguous evidence of fire *per se*, because ferns also replace vegetation massacred in other ways, but it is consistent with the global fire scenario.

[10] Besides soot and PAHs, the wildfires also yielded a fair amount of toxic gases, mostly in the form of carbon monoxide and dioxide: up to 100 and 900 ppm respectively.

Poisons and toxic waste

Ejecta bombardment and global wildfires were the most lethal aftermaths of the impact, at least on dry land. Another blow to the biosphere could have been chemical poisoning of the air, water and soil.

The idea of poison falling from the sky is not new. For centuries comets have been considered bad omens and blamed for spreading poisons and disease. In fact, when astronomers began to study the chemical make-up of comets through the technique of spectroscopy, they did find some alarming substances. At the turn of the century, the spectral detection in Halley's comet of cyanhydric acid (HCN) in particular – a most violent poison – contributed to the panic that shook the world during its close passage of 1910. Hydrogen cyanide was also detected in Kohoutek's comet in 1973.

When the Alvarez cosmic theory was published in 1980, geochemist Ken Hsü revived the issue by pointing out that the cyanhydric acid of a cometary impactor might well have poisoned the biosphere at K-T time. However, as chemist Wendy Wolbach later pointed out, the thorough vaporization of the projectile upon impact would not have preserved a molecule as fragile as HCN. The original gases of the impactor were clearly decomposed in the fireball.

On the other hand, poisoning from the bolide might have come from its non-volatile elements, namely its heavy metals. Meteorites and their parent bodies are rich in metals like nickel, chromium and cobalt, which are notoriously toxic.

Chemists Stewart Davenport and Thomas Wdowiak stressed the point by exposing radish seeds – a species well known for its resistance – to powdered meteoritic material (namely a crushed piece of the Allende meteorite, which is a carbonaceous chondrite). The plants were seen to wither away soon after germination, as their production of chlorophyll collapsed.

As it was, the Chicxulub impact must have spread a sizeable amount of toxic metals around the globe. The rate of nickel deposition, in particular, must have reached 50 grams per square meter, which equates to concentrations of 100 to 1000 ppm in the topsoil. The threshold of toxicity for most plants is around 40 ppm.

Does one have any proof of high concentrations of toxic metal in the K-T layer? The evidence is difficult to collect because nickel and similar metals are soluble in water and are readily washed away from sediments. Nonetheless, Virgil Sharpton *et al.* measured nickel concentrations of up to 40 ppm in the K-T layer of Alberta, which reaches the toxic level. Initial concentrations were probably much higher. Likewise, Huffman *et al.* have measured nickel concentrations reaching 140 ppm at the K-T horizon in cores from the South Atlantic.

Toxics from the impactor were one thing. But chemists found an even greater cause for concern in the toxic byproducts created by the impact in the atmosphere and the target rock.

The atmosphere can be the realm of a variety of harmful chemical reactions, as we have learnt at our expense from industry-produced acid rain and the seasonal depletion of the ozone layer.[11] At the time of the K-T impact, the initial blast and the ejecta fallout caused shock waves to ripple through the Earth's atmosphere, raising temperatures to over 2000 K along the shock front. At such temperatures, the gas molecules of the atmosphere ionize and recombine to form harmful nitrogen oxides (mostly NO and NO_2). One can speculate that the Chicxulub impact raised the proportion of these noxious gases to several percent locally along the most energetic shock fronts.

Nitrous oxides in the atmosphere combined with water vapor to form droplets of nitric acid (HNO_3) that rained upon the surface of the Earth. These acid rains probably reached a pH as low as 1. In the oceans, they spread essentially in the surface waters, and perhaps inhibited the formation of calcareous tests and shells that support planktonic life.

On dry land, these acid rains also contributed to the devastation of life. Acid runoff extracted soluble substances from the continental rock and soil – including harmful metals – to concentrate them in rivers and lakes. This scenario is supported by the high levels of

[11] Nitrogen oxides could have dissociated ozone in the upper atmosphere during the K-T impact, adding another threat to the already stressed biosphere. However, other chemicals and aerosols generated by the impact would have taken up the ozone's role in intercepting ultraviolet radiation, so that the extent of UV harm to the biosphere is uncertain. For all intents and purposes, harm from UV exposure was probably minor with respect to other killing mechanisms.

arsenic, selenium and antimony found in the dark sediment overlying the K-T boundary on many sites. Mercury and aluminum are expected to have leached out of the ground in vast amounts as well. And indeed, on several continents, Alan Hildebrand and William Boynton have reported anomalously high concentrations of mercury at the K-T boundary.[12]

Sulfur unleashed

Noxious oxides and acid rain were not generated by atmospheric gases alone. Vaporized rock in the ejecta also contributed significant amounts of acid gases to the atmosphere. The severity of this 'target effect' was realized as soon as Chicxulub Crater was identified and its geological setting studied by Alan Hildebrand, Kevin Pope, volcanologist Haraldur Sigurdsson, geologist Robin Brett and Yucatan expert Eugene Perry, to name but a few. All stressed that the peninsula's limestone was interbedded with thick layers of anhydrite, a mineral rich in sulfur. In the Pemex cores, outside the crater rim, anhydrite layers reach several hundred meters in thickness. These layers must have been vaporized within dozens of kilometers around ground zero, injecting close to 100 billion tons of sulfur dioxide (SO_2) into the atmosphere.[13]

Like the nitrogen oxides formed in the shock waves, the sulfur dioxide also mixed with water vapor to form acid clouds. But before it rained to the ground, the sulfuric acid might have stayed suspended several years in the stratosphere, in the form of small, shiny ice crystals. These bright sulfuric clouds could have reflected a significant amount of sunlight back into space, and contributed significantly to the global cooling that we later describe.

The sulfur kick was also accompanied by a carbon spike. Indeed,

[12] The concentrations of acid rain might have been no more severe than in industrial regions today, according to recent calculations by Owen Toon *et al.* The main difference is that their influence would have extended worldwide.

[13] This sulfuric pollution, due to the abundance of anhydrite in the Yucatan target rock, was unexpected: only 0.5 % of the Earth's surface is that rich in sulfur.

Figure 6.8. The fishing harbor of Puerto Chicxulub, on the Yucatan coast, is at ground zero of the K-T impact. The Chicxulub blast is thought to have been especially devastating because of the abundance of carbon-rich limestone and sulfur-rich anhydrite on the target site. (*Photograph: by the author.*)

the target sediment vaporized in the Yucatan also included abundant limestone, which added vast amounts of carbon dioxide to the atmosphere. Through the greenhouse effect, this carbonic gas might have promoted an episode of global warming.

A cooling effect due to sulfuric clouds and a greenhouse warming from carbonic gases are by no means contradictory: they were played out on different timescales in the aftermath of the impact and harmed the biosphere with independent blows. One can imagine the heat pulse and global wildfires singeing the Earth in a matter of hours and days, followed by a sharp cooling trend lasting months to years, due to the dust, soot and sulfur aerosols. When the opaque veil finally dissipated and the greenhouse effect kicked in, temperatures then swung to the other extreme for centuries or thousands of years.

Figure 6.9. Inland from its fishing harbor, the peaceful village of Chicxulub is surrounded by shrub land and plantations of sisal cacti. Chicxulub is a Mayan word, believed to mean 'the horns of the devil', a fitting name for an impact that raised havoc on Earth. (*Photograph: by the author.*)

The big chill

A dust cloud obscuring the face of the Earth was the first doomsday scenario put forth by Alvarez *et al.* in their 1980 paper. By scaling up the effects of great volcanic eruptions and nuclear tests, they concluded that the rock dust kicked up by the impact would have stayed suspended up to two years in the atmosphere, bringing total darkness on the planet, interrupting photosynthesis, and collapsing the food chain.

The atmospherics were explored by NASA specialists Owen Toon, J.B. Pollack, T.P. Ackerman and others who pointed out that an impact of such magnitude did not resemble volcanic and

Figure 6.10. The eruption of Rabul volcano in New Guinea, photographed in September 1994 by the Space Shuttle. The eruptive plume spread a thick blanket of ash in the atmosphere. Important ash plumes are known to lower global atmospheric temperatures by about one degree. The thick and durable ash blanket of the K-T impact is modeled to have lowered continental temperatures by several tens of degrees. (*Photograph: NASA.*)

nuclear models developed so far, but was more comparable in magnitude to the dust storms of Mars.

By analogy to Martian dust storms, Toon *et al.* estimated at first that the fine ejecta that was injected into the stratosphere spread around the Earth in less than two weeks, due to the strong winds generated by the temperature differences between dusty and clear air. They now believe that the initial blast distributed the dust even faster, by way of the global fireball itself.

Once the dust was globally distributed, its suspension time in the atmosphere depended on a number of factors, such as the rate of coagulation of the dust specks into larger grains (faster rates of settling) and rain washout of the dust in the lower atmosphere. According to the models of Toon *et al.*, micrometric particles would have survived in suspension for up to several months in the stratosphere. Once they crossed down into the troposphere (under 10 km), they were precipitated by rain to the ground in a matter of days.[14]

As long as they resided in the stratosphere, the micrometric particles intercepted the sunlight so efficiently that pitch darkness prevailed for about two months over the entire world, followed by at least six extra months of twilight conditions, under the minimum light levels required for the photosynthesis of plants. This was enough to massacre plankton in the oceans and lead to a collapse of the marine food chain, as was rightly predicted by Alvarez *et al.* in their 1980 paper.

More dramatic models, involving longer blackout times, take into account the amount of soot added from global wildfires and the sulfur aerosols we described in the target effect.

In the weeks that followed the impact, the unending night did more than simply turn off photosynthesis: it also caused temperatures to plummet. Toon *et al.* initially estimated that the global temperature should have dropped by at least five degrees Celsius, and Pope *et al.* suggest a longer and steeper ten degree drop, since sulfur aerosols might have intercepted 80 % of the sunlight for up to ten years. Continental interiors were certainly more affected than oceanic regions, and temperature drops could have reached 40 degrees far inland, a disagreeable switch from balmy averages of +20 °C to a freezing hell of −20 °C.

Sea temperatures would have been less affected than air by the frigid night, in view of the ocean's large thermal inertia. The temperature of surface waters probably dropped no more than two to three degrees, which is nonetheless a harmful temperature jump for many a species.

Does one have any proof of a planet wide temperature drop at the K-T boundary? It does not appear so. In marine fossils, a good

[14] The washout was all the more efficient in that a deluge probably raged for weeks in many parts of the world, due to the amount of water vapor that the impact also kicked up into the atmosphere.

temperature indicator is provided by the oxygen in their limestone: the element's two different isotopes, O-16 and O-18, are present in ocean water and incorporated into living shells in a ratio that is closely related to water temperature. At the boundary, one does indeed observe a shift in the oxygen ratio of the fossils, but it points to a rise and not to a fall of the temperature, as we shall see in the following section. This is no great wonder: a cooling shift lasting only two or three years would not have had the time to get incorporated into the meager sedimentary record of the devastated biosphere.[15] Only the subsequent, lasting global warming would have had that opportunity.

The greenhouse effect

Contrary to the evanescent big chill, the warming trend left a durable mark in the sedimentary record. The O-18/O-16 ratio of marine fossils shows a substantial increase above the K-T boundary, marking an average temperature surge of at least 5 degrees in the water (and perhaps as much as 15 degrees in the air) over a period of thousands of years.

From a theoretical standpoint, such a temperature rise was expected of the greenhouse effect that an impact of the size and target nature of Chicxulub would set in motion.

The greenhouse effect is the heat trap caused by gases that let sunlight through to the ground, but do not let infrared radiation from the heated Earth radiate back into space. The result is an increase in the atmosphere's temperature, until a new balance is struck. Carbon dioxide is often cited as a major greenhouse gas, in view of its heat retaining capacity and relative abundance in the Earth's atmosphere. Water vapor is also an efficient greenhouse gas, as are methane and several of the nitrogen oxides.

[15] Kenneth Hsü and others believe that the cooling effect could have played for centuries, if snow precipitated during the initial phase of the chill, heightening the reflectance of the Earth and sustaining the cooling trend. One clue is the abundant iceberg-rafted 'dropstones' in the sediments just above the K-T boundary, off the Spitsbergen Islands (especially when one knows that there were no polar ice caps in the balmy days of the Late Cretaceous).

In the case of the Chicxulub impact, a significant amount of water vapor shot out of the ocean: along with the nitric acid generated by the shock waves, it probably was active in trapping heat in the early aftermath of the impact, before washing out of the atmosphere.

But it was carbon dioxide that was by far the most plentiful and active greenhouse gas released by the impact: vaporized out of the limestone target rock, carbonic gas could have reached an instant production of 1000 billion tons. This already amounts to a quasi doubling of the atmospheric CO_2 concentration. The burning of the biosphere added more carbon dioxide to the balance, raising the concentration one more notch. Finally, the disappearance of the marine plankton eliminated the main carbon sink in the oceans. On the contrary, because water acidity increased, marine calcium carbonate dissolved and returned even more carbon dioxide to the atmosphere.

All in all, the CO_2 concentration in the atmosphere must have increased by a factor of four or five in the weeks and months following the K-T event. Such a jump in concentration would translate in theory into an atmospheric temperature rise of five to ten degrees. These calculations confirm the five to ten degrees estimate derived from isotope ratios in the sediment.

The global warming probably lasted for thousands or tens of thousands of years, until the biosphere recovered sufficiently to spread photosynthesis in the oceans again, and pump carbon dioxide out of the atmosphere and back into the waters.

Selectivity of the extinctions

In the impact scenario, one then has the combined effect of the prodigious energy release, giant tsunamis and global wildfires, noxious gases and poisonous metals, total obscurity and deep freeze, acid rain, and global warming. These lethal mechanisms played out differently on various ecosystems and species.

The challenge now for paleontologists is to confront the impact model with the pattern of extinctions – constantly refined by new studies – in the End-Cretaceous biota. Does our knowledge of which species went extinct and which survived square with the proposed mechanisms of doom? What sort of selection was at play?

At first view, the record of the End-Cretaceous mass extinction shows the kind of ecosystem collapse that one would expect from long-lasting obscurity, due to the dust and soot clouds: a massacre of photosynthesis-driven algae, plankton and plants, and the resulting collapse of the primary food chains in the oceans and on dry land.

The mass killing at the ocean surface manifestly struck all forms of plankton and other floating and swimming microorganisms, and notably the *foraminifera* – those relatively large (millimeter-sized) protozoans with calcareous tests. Only a few 'dwarf' species among those that were widely distributed in all oceans – and thus were most adapted to changes in water temperature – made it through the catastrophe with enough survivors to eventually repopulate the oceans when conditions improved.

As for those planktonic species that build their tests and shells out of hard silica rather than calcium carbonate, such as the *diatoms* and *radiolarians*, the cutdown in their ranks was apparently short lived and they recovered quite soon after the K-T boundary. Building their hard parts out of acid-resistant silica allowed them to better cope with the adverse environment. Perhaps as well, the increased availability of silica from vigorous erosion on land helped the spared species proliferate anew.

All these planktonic species, which constitute the base of the ocean food chain, live in the surface waters lit by the Sun – the photic zone. Marine microorganisms living on the sea bottom had a much lower extinction rate than those at the surface: few of their species disappeared at the K-T boundary. Obviously, these *benthic* species did not rely on photosynthesis for survival, and they were also insulated under tens of meters of water from any heat blast, drastic cooling or acid rain. On the contrary, they even got an increased supply of food, in the form of dead organic matter trickling down from the devastated surface.

Terrestrial ecosystems also show conspicuous trends in the fossil record. From the crash in pollen and its replacement by resistant fern spores (especially in North America and Asia), we know that the vegetation was massacred and photosynthesis virtually turned off. In fact, carbon overlying the K-T boundary seems to come largely from bacterial activity in the devastated soil, rather than from photosynthesis.

With the base of the food chain suddenly cut off, higher organisms were condemned. Without plants, the herbivores could not survive; and without herbivores to feed on, the carnivores were doomed in turn. This domino effect was amplified by the more stringent metabolism and reproduction requirements of large-sized animals.

Indeed, large species require more food per kilogram of body weight than smaller ones, an obvious disadvantage when food is scarce. Moreover, because large-sized species have fewer individuals than smaller species (there are fewer elephants than, say, frogs), losing a given percentage of a population might have more severe repercussions. For instance, a dinosaur species counting 2000 members and suffering a 90% kill will have much trouble recovering with a pool of only 200 survivors – especially in times of stress when sexual reproduction is inhibited. On the other hand, a dragonfly species of two million individuals losing 90% of its stock will still have 200 000 survivors after the disaster, a population density which gives the species a much higher chance of reproduction and recovery.

Starvation, failure in reproduction or both: it is a fact that no land species over 25 kilograms in weight survived the K-T crisis. Among the smaller amphibians, reptiles and mammals, there were a number of species that pulled through the crisis, however, and paleontologists are now attempting to discover which factors, other than sheer luck, insured their survival.

The survivors

Besides small body weight and large number of individuals, one can imagine a handful of other factors that might have played a selective role in the survival of species across the K-T boundary.

One such factor is the method and timing of reproduction and development. For example, seeds, cysts and other resistant DNA 'time capsules' can ride out long periods of adverse conditions –

months to years – before hatching new individuals. This time-delay 'strategy' would have allowed certain species to survive the freezing obscurity that spread over the Earth in the aftermath of the impact.

Moreover, those species which hatch fully constituted individuals, rather than fragile and transient larvae, would also have been advantaged. This was especially true in the oceans where the fragile larval stages tend to live in surface waters – one of the environments most devastated by the impact (marred by obscurity, acidity and temperature fluctuations). Fully formed hatchlings, in contrast, were capable of rapidly migrating to deeper, less disturbed waters.

A second factor of selectivity among species might have been the feeding strategy, namely its place in the food chain. This might explain the astonishing survival rate of freshwater species – be they fish, amphibians or reptiles.[16] The ecosystem of rivers and lakes is indeed very different from the land and marine systems. Whereas the latter are based on primary food production (photosynthetic algae and plants), freshwater species feed mainly on organic waste washed and blown in from the surrounding banks. Hence, the abrupt cut-off of photosynthesis during the K-T crisis did not directly nor immediately concern the freshwater habitat. On the contrary, the great kill that devastated the surrounding land might well have provided a food bonanza for freshwater animals, lasting several months to several years.

The high survival rate of freshwater species holds at least one contradiction. Acid rain should have severely affected lakes and rivers, the runoff dumping leached, toxic metals into the streams, along with the sulfuric and carbonic acid itself. What could possibly have shielded freshwater species from this aggression?

One can speculate that those rivers and lakes set in limestone terrain were protected by the alkaline chemistry of their rock beds, which acted as a buffer to neutralize the acidity of the inflowing rain and runoff waters. Freshwater species would have survived in these buffered oases. Another possibility is that there was simply

[16] Paleontologists Peter Sheehan and David Fastovsky took a census of surviving species at the K-T boundary of Hells Creek, in eastern Montana, and found a 66 % survivorship among freshwater vertebrates, as opposed to 10 % for land-dwelling vertebrates.

Figure 6.11. Only a quarter of all species on land and in the oceans survived the K-T impact. Mammals that could ride out the temperature extremes in deep burrows and scurry for scarce food in the lasting obscurity were advantaged. Half of the mammal species survived the K-T catastrophy. (*Photograph: by the author, Tucson Desert Museum.*)

not that severe an acid rain pulse at K-T time, and that it was not a major cause of extinction.

A third scenario for survival concerns us directly, since it probably affected our mammalian ancestry: the 'bunker effect'. Sea-bottom species were protected by the water blanket, as were freshwater amphibians capable of diving to the bottom of rivers and ponds. Small mammals living in burrows might likewise have been insulated from the temperature extremes of the impact's heat flash[17] and the subsequent deep freeze brought about by the dust and soot blackout (burrows deeper than a foot should have stayed above freezing temperatures).

Burrowing mammals were further protected by three other

[17] A modern day example is the eruption of the Montagne Pelée volcano in 1902, which devastated the town of Saint-Pierre-de-la-Martinique: out of the population of 30 000 that was massacred by the 700 K pyroclastic flow (a ground surge of hot ash and gas), the only two survivors were a shoemaker hiding out in his cellar and a petty thief locked up in an underground prison cell.

strong points. Firstly, they were mostly warm-blooded animals, which helped them function through the cold 'impact winter'. Secondly, they were mostly nocturnal animals, leaving their burrows and feeding at night. A blackout lasting several months was no tragedy to them. Whereas the few dinosaurs that survived the initial heat pulse were fumbling around in the everlasting night, bumping into charred trees as they searched for food, nocturnal mammals were at home in the dark as they scurried for scraps.

This brings us to the third and perhaps most important factor. Most of the End-Cretaceous mammals were omnivores that did not depend on a specific food type but fed on whatever they found – seeds and roots, insects and carrion. For these species with an unspecialized and 'opportunistic' diet, the crash of the food chain had less of an effect than on herbivores and carnivores that depended on specific sources of food.

These are tentative scenarios at best. The ball is now in the court of paleontologists, who continue to refine the record of victims and survivors across the K-T boundary, to check out the various mechanisms of extinction and survival that have been proposed.

7

Impacts and other extinctions

The world took a long time to recover from the End-Cretaceous mass extinction. Not only did the Chicxulub impact wipe out the vast majority of species on Earth, but it also massacred the ranks of the few surviving species, so that the planet was literally devastated. In the oceans this spectacular collapse of the ecosystem lasted for hundreds of thousands of years, as is readily apparent from the sedimentary record: the strata that mark the dawn of the Tertiary era are depleted in planktonic shells and other fossils. The contrast is visible to the naked eye at a number of localities on Earth, where the limestones and marls above the K-T clay are condensed and dark, impoverished in carbonates and rich instead in detrital particles. Only after tens of centimeters – even several meters in some cases – do the layers brighten up in color and resume a normal, fossil-rich appearance.

One can thus picture empty oceans that remained sterile long after the impact, a meager trickle of plankton sustaining an extremely reduced population of marine animals. Ken Hsü, in his analysis of carbon collapse after mass extinctions, coined the expression 'Strangelove oceans' to describe the utter devastation that prevailed in the waters, in reference to Stanley Kubrick's cult film about nuclear war on Earth.

Based on the sedimentary record, micropaleontologists estimate that the world oceans took at least 500 000 years to return to normal levels of biological activity, and more than a million years in those areas most affected by the catastrophe. At Gubbio, for example, planktonic richness and diversity became fully reestablished after an estimated 1.4 million years.

Because the End-Cretaceous catastrophe is relatively recent in geological history and has been studied at a number of well preserved localities on Earth, it has been possible to brush a detailed portrait of the biological crash, identify its postulated causes, and trace the slow recovery of the biosphere in the aftermath of the event. It is now timely to broaden the issue and tackle the problem of other mass extinctions on record: could they too have been caused by cosmic impacts?

Towards a general theory of impacts

Starting from the assumption that a 100 million-megaton impact, leaving a 180-km-wide scar on Earth, was responsible for the extinction of 70% of all species on the planet, one can wonder at what smaller threshold level an impact has any sort of global killing effect at all and will lead to a recognizable extinction pattern in the fossil record. This attempt at a 'unified theory' was first broached by Michael Rampino in the mid-eighties (see figure 7.1), and has gained momentum since the confirmation of the cosmic nature of the K–T crisis.

The issue is undoubtedly complex. We have seen that two impacts of similar magnitude might have different effects, depending on the angle of impact and the fragmentation of the bolide (which controls the coupling of the energy with the atmosphere); the latitude of the hit (which has a bearing on the atmospheric and marine circulation of the perturbations); the water cover on the site (impact on land or at sea); and the nature of the target rock (crystalline basement or sediments rich in carbon and sulfur). One can add to this list the degree of stability of the biosphere – i.e. to what extent the Earth is already stressed by other factors before the fatal blow.

If one leaves aside these many levels of complexity, one can begin by defining a ballpark 'threshold of harm' by asking the following question: how large must an impact be to kick up sufficient amounts of ejecta into the atmosphere to cause a worldwide heat pulse capable of causing global wildfires, followed by a long term

Figure 7.1. Besides the K-T layer, researchers like Michael Rampino of New York University have studied other extinction boundaries in search of impact clues. Perhaps as many as half of all mass extinctions of the fossil record are due to asteroidal and cometary impacts. (*Photograph: courtesy of Michael Rampino.*)

darkening and cooling of the atmosphere? Computer models put the figure at 10^{23} joules (10 million megatons of TNT) – a blast strong enough to excavate a 150-km-wide crater.

This theoretical figure finds some justification in the fact that the impact frequency of 150 km (or larger) craters on Earth is of one event every 100 million years on average,[1] which is indeed the average frequency of the great mass extinctions on record (five in the last 500 million years). One can also reverse the reasoning: with respect to the more numerous, smaller extinctions on record, what smaller impact size might possibly have caused those?

According to the statistics compiled by paleontologist Jack Sepkoski, there could be as many as 25 mass extinctions (the five large ones and 20 minor ones) in the marine fossil record of the past 550 million years. This equates to a frequency of about one crisis every

[1] The frequency of impacts on Earth and their size distribution (the smaller, the more frequent) can be estimated from the inventory of craters on the Earth and Moon, as well as from the observed flux of asteroids and comets crossing the Earth's orbit.

22 million years, on average. If we then look at the distribution curve of craters in the record, the events that occur every 22 million years or so would be 'mid range' impacts, leaving craters 80 to 100 km in diameter. From this frequency match, and assuming of course that all mass extinctions are due to impacts, one can then suggest that an impact triggers a mass extinction at crater sizes around 80 to 100 km, and that the larger threshold of 150 km corresponds to the greater mass extinctions on record (the Big Five).

Recent theoretical work by Toon *et al.* (1994) does confirm that the threshold impact energy for causing a thick planetary dust cloud is around 10 million megatons of TNT (corresponding to craters of 100 km), but slightly higher for global wildfires (100 million megatons TNT), which are spread only by impacts the size of Chicxulub (craters greater than 150 km).

Looking at the record

One can always speculate over theoretical impacts and mass extinctions, but the only way to prove anything is to examine the sedimentary record at each extinction level and search for clues of impact: iridium and other geochemical anomalies, shocked minerals, tektites and ablation spinels, and of course impact craters of the right age.

By no means is this global inventory an easy task. The older the extinctions, the more difficult it is to find well preserved strata that bear unaltered geochemical and mineralogical evidence. Radiometric dating loses precision as one probes deeper in time, and distant sites around the world are difficult to correlate one to another, disagreements arising as to their temporal coincidence. In such a widely scattered puzzle, deciding that a geochemical anomaly is consistent worldwide, rather than a local oddity, becomes a daunting task. But as dating techniques improve and more scientists look at rock strata with the new understanding that events can be sudden and global, more telling horizons like the K–T layer are bound to be discovered.

Extinction specialists are beginning to sum up the evidence, as have Digby McLaren and Wayne Goodfellow in 1990, and Michael Rampino and Bruce Haggerty in 1995. Following in their footsteps, let us then take a look at the record of life on Earth and at the many sharp events and mass extinctions that steered the course of evolution through time.

The appearance of life on this planet can be traced back to the oldest rocks still preserved at the surface – fossils of bacteria and blue-green algae that became pervasive in the world oceans about 3.6 billion years ago. These primitive single cells did not evolve further for over a billion years until, around two billion years ago, a major improvement took place with the surge of photosynthesis and a drastic release of oxygen in the atmosphere. A second revolution occurred soon thereafter, about 1.8 billion years ago, with the appearance of the first multicelled algae – a major step towards complexity. After these two momentous improvements, life did little more than rest on its laurels – or should we say seaweed – for the next billion years.

There is an odd coincidence here. The two largest astroblemes discovered so far on Earth – Vredefort (300 km) in South Africa and Sudbury (250 km) in Ontario – are dated at 2.02 and 1.85 billion years. Undoubtedly these giant impacts massacred the algal community and led to grave perturbations of the environment. Could it be, then, that in the ensuing biotic 'reset', the decimated algae found the right opportunities for a new start and a successful jump in complexity?

The Cambrian explosion

The early era of simple life on Earth – the Precambrian – thus holds some fascinating secrets still waiting to be solved. But it is with the sudden 'explosion' of complex lifeforms in the Cambrian period, around 550 million years ago, that our 'high fidelity' record of evolution truly begins.

Why life suddenly switched gears from colonies of simple algae and bacteria to an extraordinary palette of complex animals is still a mystery, but our understanding of this crucial Precambrian–Cambrian transition is progressing by leaps and bounds.

There is no consensus yet as to where exactly one should place the boundary – a prime example of the difficulty of correlating fossil-rich strata from different continents. From the fossil-rich Burgess Shales of the Canadian Rockies, paleontologists have traveled to the Yangtze Gorges of China, the thick strata of north-eastern Siberia and the rugged outcrops of Namibia, not to mention the early Ediacara fossils of Australia, unraveling a story of extra-ordinary biological competition and diversification. In this frenzy of early Cambrian creativity, which set up the blueprints of just about every branch of life on the tree of evolution, there are at least two blatant signs of cosmic intervention.

The first is a giant impact scar in Australia – Lake Acraman Crater, due west of Adelaide – which is estimated to be anywhere from 100 to 160 km in diameter. Its ejecta layer, rich in iridium and shocked quartz, extends 300 km east of the impact site into a sedimentary basin where it underlies the famous Ediacara fossil assemblage – one of the earliest pulses of complex life that immediately precedes the Cambrian explosion. Again, the possibility that a major impact wiped clean the biological slate and allowed new lifeforms to evolve must seriously be considered.

There is a second event at the Precambrian–Cambrian boundary that hints at another major impact affecting the biosphere. It was first described by Ken Hsü at the Yangtze Gorge of southwest China, where it takes the form of an abrupt shift in carbon isotopic ratios, coupled with a marked increase in iridium (3 to 4 ppb), osmium and gold. The negative shift in carbon values, which is also found in contemporary sediments of Siberia, North Africa and the Lesser Himalayas, is interpreted to represent a catastrophic destruction of the biosphere – a blatant case of 'Strangelove ocean'. Immediately above this sharp boundary, dated at 527 million years, the first trilobite fossils appear, marking the definitive blossoming of complex life in the oceans. No impact scar has yet been discovered for this most important event in the history of life, but one can guess that it hit the very scene where the drama of evolution was being played out: the open sea.

The end of the Ordovician

The blossoming of life proceeded steadily for 100 million years through the Cambrian and Ordovician periods, with only minor extinction spikes, until a spectacular event marked the end of the Ordovician: the Ashgillian event, dated at 438 million years.

First of the five great mass extinctions of the geological record, the Ashgillian or End-Ordovician event consists of a complex global crash of the marine ecosystem (life had barely set foot on dry land, so there is no terrestrial record).

A careful look at the marine fossils shows that the crisis occurred in three steps, within less than a million years. The extinction apparently began among tropical species of plankton, echinoderms (the ancestors of starfish and sea urchins), trilobites and armored fish; it then hit coral and brachiopod species (shellfish) before ending with a sudden crisis that affected both surface and bottom waters.

The final crisis is marked by a swing in carbon isotopic ratios, indicating a sudden biomass kill and a mixing of surface and bottom waters. The carbon trend is paralleled by a swing in oxygen isotopes that points to rising temperatures and a greenhouse effect.

The consensus among geologists and paleontologists is that the End-Ordovician double-staged extinction is due to the onset of continental glaciation, draining continental seas and upsetting oceanic circulation, followed after a few hundreds of thousands of years by the abrupt termination of that glacial period and a renewed upheaval of the biosphere.

If one wants to play devil's advocate, one can wonder if the onset of glaciation, its termination, or both, were not due to cosmic impacts. And indeed, there does happen to be a slight anomaly in iridium at the End-Ordovician boundary, at the horizon of the terminal biomass kill, reported in samples from Alaska, Canada, Scotland and China. Although iridium is not accompanied by other rare metals in cosmic proportions, the fact that it accompanies a biomass crash and a temperature rise does parallel the pattern observed at the K-T boundary.

However, apart from this modest iridium anomaly, the End-

Ordovician mass extinction shows no direct evidence of impact: no shocked quartz, ablation spinels or tektites. The End-Ordovician also lacks a major impact crater. Actually, the entire time interval from 500 to 400 million years lacks impact craters: only three small occurrences were known in the mid-nineties.[2] Apparently, there is a poor record of impacts at the time, perhaps because of the scarcity of strata of the right age, and their frequent folding and alteration that do not easily preserve astroblemes.

A puzzling coincidence, however, is that the three impacts on record in the 500–400 million year bracket converge on the age of the End-Ordovician mass extinction (438 million years): Strangways (24 km) in Australia, Pilot Lake (6 km) and Brent Crater (4 km) in Canada. Pilot Lake, notably, is precisely dated at 440 ± 2 million years.

Of course, these diminutive impacts could not themselves have had any effect on the biosphere, but one can legitimately wonder if they were not part of an asteroid or comet salvo, in which case some larger objects might have hit the ocean and caused the noted biomass kill. Such an ocean impact would have left little evidence anyway, since all Ordovician ocean crust has today vanished in the trenches of subduction zones, carrying away with it hypothetical astroblemes and ejecta blankets.

The end of the Devonian

While the first great mass extinction – the End-Ordovician – carries only modest hints of an impact, the second great mass extinction of the fossil record – the End-Devonian – is riddled with cosmic clues, not only of one, but of several impacts.

Since the previous crisis, life had proceeded steadily through the Silurian and Devonian periods, increasing in abundance and diversity. Plants had crept out of the sea to colonize the dry land, soon to be followed by the first amphibian animals. Towards the

[2] The search continues: new craters of End-Ordovician age have recently been reported in Scandinavia.

end of the Devonian, however, all hell broke loose in the biosphere.

A major mass extinction occurs between the two last 'stages' (sedimentary sequences) of the Devonian – the Frasnian and the Famennian – and is thus known as the F-F boundary event, dated at 367 million years. The change is sudden in sediments all over the world, where fossil-packed limestones are abruptly replaced by sterile black shales on many sites. Coral reefs, which had been growing and diversifying at the end of the Devonian, appear to collapse rapidly just before the boundary, perhaps due to some pre-extinction stress.[3] At the boundary itself, there is a mass killing of the marine ecosystem, with a crash of the plankton supply and the inevitable collapse of the food chain, leading to extinctions of many species of trilobites, sponges and armored fish – the last of which are exterminated. On dry land, plants also underwent a major shake-up at the boundary, marked by a major change in the spore record.

The crisis is so abrupt and devastating – the early Famennian sediments, above the boundary, are notoriously poor in fossils – that paleontologist Digby McLaren proposed an impact scenario for the F-F boundary as early as 1970, ten years before the Alvarez team found evidence for an impact at the K-T boundary.

Similarly to the K-T, the F-F boundary often takes the form of a thin clay, showing a net carbon isotope anomaly pointing to a biomass massacre and a 'Strangelove ocean' of sterile waters. There is also an oxygen isotope anomaly pointing to a greenhouse effect. What, then, is the evidence of impact?

There is at least one level of iridium enrichment in the vicinity of the F-F boundary. Discovered in Australia, Europe and China, the modest iridium concentration (0.2 to 0.3 ppm – a 10 times increase over background values) was first attributed to metal-selective metabolism of bacteria, which could have precipitated iridium out of sea water and into the clay, but the fact is that the waters would have had to be anomalously enriched in iridium anyway. Moreover, the F-F iridium shows a meteoritic chemical ratio with the other rare metal ruthenium, in both New York State and Australia.

[3] It appears that a greenhouse effect set in about 500 000 years before the F-F boundary, perhaps triggered by a preliminary impact, or by some other cause.

Even more convincing of a major impact at the F-F boundary are the vitreous microtektites discovered in China, a couple of meters above the F-F boundary, suggesting an impact one to two million years after the initial crisis.[4]

Hints of multiple impacts do not stop at this double-barreled blast just at, and above, the F-F boundary. Only a few more meters of geological time above those two layers lies the final blow to the doomed End-Devonian world, a last pulse of extinction marked by another biomass kill (swing in the carbon isotope ratios), accompanied by the extinction of almost all ammonoids – a family of graceful shellfish – and of many species of trilobites. This third massacre is the official passage from the Devonian into the Carboniferous period, 360 million years ago, known as the D-C boundary. In North America, the D-C boundary is marked by several closely spaced anomalies of iridium.

The End-Devonian sequence of catastrophes is also *preceded* by evidence of impact. Indeed, indirect marks of impacts in the End-Devonian are present in the form of giant tsunami deposits: one of these is the Alamo breccia of Nevada, estimated to have formed less than three million years before the F-F boundary.[5]

Studied by C.A. Sandberg and John Warme, the Alamo breccia is an unusual strata resulting from the chaotic collapse of a huge slice of sediments off the shores of Nevada's ancient sea. The giant submarine slide cut out hundreds of cubic kilometers of reef limestone and beachline, slumping it downslope in a debris flow 70 meters thick (as high as a 15 storey building), spread out over 4000 km^2 – the area of a state like Rhode Island. But, most importantly, the Alamo breccia bears iridium and shocked quartz. The impact nature of the shocked quartz is indisputable, according to a careful study by French physicists Hugues Leroux and Jean-Claude Doukhan of Lille University.

[4] The Chinese microtektites are submillimeter glassy droplets mostly composed of silica. They bear inclusions of *lechatelierite* glass, a high temperature form of silica which is key evidence for an impact origin (chapter 4). Other microspherules have been reported at the F-F boundary in Belgium, but they are probably man-made particles from industrial activity in the area.

[5] Chaotic deposits are common in the End-Devonian strata of North America, Europe, Australia and China. They were initially interpreted to be the deposits of great storms before being proposed as tsunami deposits from impacts at sea.

A strong impact at sea, close to North America, thus occurred less than three million years prior to the F-F boundary, perhaps disturbing the climate shortly before the End-Devonian crisis. The crisis was then played out in three acts, as we just saw, dated approximately at 367 (F-F event), 365 (second F-F event of China) and 360 million years (D-C event).

Besides the geochemical and mineralogical evidence of impact at the extinction boundaries, impact craters of the right age seem to abound. Richard Grieve lists six astroblemes that cluster around the End-Devonian time period, including the small craters of Kaluga (Russia), Lac La Moinerie (Quebec), Misarai (Lithuania) and Flynn Creek (Tennessee) – all under 15 km in size – and more importantly the Siljan astrobleme of Sweden (52 km) and the Charlevoix astrobleme of Quebec (46 km), respectively dated at 368 ±1 million years (exactly matching the F-F boundary) and 360 ±25 million years.

Although both Siljan and Charlevoix (not to mention the diminutive other craters) are theoretically too small to account for a global mass killing of the biosphere, their combined effect, if they occurred as a simultaneous hit or as part of a larger salvo, should not be discounted.

K. Wang and coauthors also suggest Taihu Lake near Shanghai as an additional F-F astrobleme: an estimated 70 km in diameter, it is said to be associated with shock metamorphism in End-Devonian sediment and would be a prime candidate for the microtektites found in China, at the level of the second mass killing.

Aside from the established and prospective continental astroblemes, other impacts apparently hit the oceans – as we know from the tsunami deposits – and these could have been even greater in size. The End-Devonian crisis, then, shows all the symptoms of an impact salvo, which caused a cascade of biomass killings and multiple pulses of mass extinction.[6]

[6] Besides the fact that impacts might be bunched within several thousand years, one impacting body may also split up into several pieces, showering the Earth along a 'firing line' over a time interval of minutes, hours or days (see chapter 8). Such a chain of impact craters has long been known in the USA and recently reviewed by Michael Rampino and Tyler Volk. It contains eight craters (5 to 17 km in size) aligned over a distance of 700 km and apparently 310 to 330 million years in age. Another chain has recently been discovered in Chad: the four aligned craters, each about 10 km in size, have been detected on Landsat and Space Shuttle imagery and their age is estimated at about 360 million years, coincident with the End-Devonian crisis.

The End-Devonian impact shower, if it is confirmed, could be either cometary in nature, with a salvo of objects flying in from the perturbed outer reaches of the Solar System, or the doing of an asteroid shake-up closer to home, leading to a surge of Earth-crossing bodies. Be it as it may, the End-Devonian is runner-up to the End-Cretaceous in terms of evidence of impact in a great mass extinction.

The end of the Permian

Chronologically the third of the great mass extinctions of the fossil record, 250 million years ago, the End-Permian crisis is by far the most severe. At this complex transition that marks the passage from the Permian to the Triassic period (and thus called the P-TR boundary), over 90% of all species – marine and terrestrial – are eradicated from the fossil record, a much larger figure than the 70% extinction rates of the earlier End-Ordovician and later End-Cretaceous crises.

The marine realm in particular was severely hit, with a 95% extinction rate of its surface-dwelling *foraminifera* plankton species, and a near equal massacre of its bottom-dwelling organisms like the fixed bryozoans and corals, shellfish like brachiopods, and trilobites – a group which had survived the two preceding mass extinctions but not this one: its last species disappear at the boundary.

On land, where life had reached a record diversity, the End-Permian crisis wiped out the distinctive flora of the period to replace it by ferns and fungi, before a new assemblage of plants slowly emerged from the 'ashes'. The majority of insect species were likewise exterminated – the only severe mass extinction of insects in the fossil record. Amphibian vertebrates lost over 80% of their species, and the successful new land-dwelling group – the reptiles – lost over 90% of their species and virtually had to start over from scratch after the extinction.

Survivors set a new pattern of evolution. On land, for example, the spared *Lystrosaurus* vertebrate was one of the founders of the modern mammal group, which evolved alongside new families of reptiles and the emerging dinosaur group. Equally important changes took place in the oceans: the predominantly bottom-dwelling fauna was replaced by a more varied ecosystem with more swimming predators like fish, squid and ammonite-like creatures. In the air, most insects with fixed wings (like the dragonflies) were replaced by new models of flying machines – insects with deployable and retractable wings.

The great End-Permian extinction, which marked the end of an era and a reorientation of evolution, was long thought to have spanned millions of years, but careful studies of the fossil record show it to have been much more short-lived than initially thought – certainly less than a million years with a final mass killing event,[7] represented worldwide by a boundary containing a strong carbon isotope anomaly (reflecting a radical drop in biomass), accompanied by an oxygen isotope anomaly (suggesting global warming), and followed by evidence of anoxia (low dissolved oxygen in the waters and high carbon dioxide, both causes of animal asphyxiation).

On land, an abrupt catastrophe is also indicated by a sharp horizon in the sediments, where the pollen of Permian plants is replaced by a global spike of fungal spores. Fungi – mushrooms and other non-chlorophyllian molds – are known to proliferate in disturbed ecosystems where there is an abundant supply of dead biomass. Moreover, it should be noted that in sediment cores collected in Israel, the P-TR boundary also contains a large amount of carbonized plant debris.

Many scenarios have been proposed for the P-TR mass extinctions: sluggish oceanic circulation leading to stratified waters with low dissolved oxygen, high carbon dioxide and marine life asphyxiation; continental ice caps and global cooling, accompanied by a damaging drop of sea level; and massive, trap-style volcanism in Siberia and China, likewise altering the climate.

The Siberian Traps, in particular, are noteworthy: they represent a very large volume of erupted material (two to three million cubic kilometers), straddle the Permian–Triassic boundary, and appear to be shorter-lived and more violent than were the Deccan Traps

[7] Perhaps there were two closely spaced mass killing events rather than one, since there appears to be a precursory swing in carbon isotopic ratios four meters (about 100 000 years) before the P-TR boundary.

at the K-T boundary (see chapter 3). If confirmed, the briefness of the Siberian Traps – perhaps less than a million years – and their high content of explosive ash layers (up to 20% of the strata) would be highly unusual. Berkeley geologist Paul Renne and coauthors suggest that the extraordinary magmatic pulse (a hot-spot plume) uplifted the northern edge of the continental land mass,[8] resulting in ice accumulation on the ensuing swell, and facilitated the rise of vast quantities of sulfur and carbon dioxide into the stratosphere.

And what about an impact scenario? The P-Tr boundary has long been thought to hold no evidence of impact, but this view is rapidly changing. First of all, there are two weak iridium anomalies a few meters apart – found in India, China and Europe – that coincide with the boundary itself and with the precursor carbon isotope anomaly that occurs roughly 100 000 years earlier. Although the iridium data is controversial and inconclusive on its own, layers of spherules that might be impact microtektites have been identified in conjunction with the iridium at P-Tr sites in China by Chinese geologist Xu Dao-Yi.

Last but not least, in 1996, Gregory Retallack of the University of Oregon reported shocked quartz at the P-Tr boundary in Australia and Antarctica. The shocked crystals are small and could come from marine sediments rather than continental basement rock: in that case the impact occurred at sea, off the shores of Australia and Antarctica, in the ancient South Pacific. Already in the past, Michael Rampino had pointed out two large circular structures on the Falkland oceanic plateau, each 300 km wide, the rocks of which show signs of an important metamorphic event – a high pressure or temperature alteration – 250 million years ago. On land, one should also note the Araguinha astrobleme of Brazil, 40 km wide, which is dated at 247 ± 5 million years.

Carbon dioxide buildup in stagnant oceans? Volcanism and continental glaciation? Giant oceanic impacts? One is confronted with a

[8] At the time, the continents were still regrouped in one landmass called Pangaea, with the Americas glued to Eurasia and Africa – itself glued to Antarctica, India and Australia. Oceanic waters occupied the entire opposite hemisphere, and penetrated in wedge-like fashion between Africa and Eurasia, forming the Tethys Sea (ancestor to the Mediterranean).

complex riddle in which it is easy to confuse cause and effect. For example, Michael Rampino pointed out that the P–TR boundary in the deep sea off Japan shows that the massacre of the biomass (namely radiolarian plankton) *precedes*, rather than follows, the onset of oxygen-poor conditions. The latter would then be a consequence, rather than a cause of extinction.

Similarly, one can question the significance of the Siberian Traps. Apparently, there was some precursor magmatic activity in Siberia two to three million years before the mass extinction, so that there is no doubt that the province was a budding hot spot at the time. At any one time, one must recall, anywhere from 10 to 20% of the Earth's surface is characterized by some form of ongoing hot spot activity (at present, about 50 hot spots are active on Earth).

One can speculate, then, that the extreme briefness and unusual pyroclastic nature of the Siberian Traps might have been triggered by a violent impact, augmenting and releasing the energy of an already formed plume, close to the Earth's surface. We have seen in chapter 3 that a 180-km impact crater like Chicxulub would not provide enough energy to initiate or amplify eruptions, be it at the impact site or at the antipodes. However, a larger impact might well do the trick, and the odds of hitting a hot spot are real: given a 20% spread of hot spots at the Earth's surface, and the fact that an impact focuses seismic energy in two different areas (point zero and the antipodal point), the chance of an impact hitting a hot spot or its antipodal point are 40% – virtually one in two!

In this light, the fact that shocked quartz and circular underwater structures are found in the vicinity of the South Pacific, close to the antipode of the Siberian Traps, is perhaps not a coincidence. And of course, a large impact at sea would overturn the stratified waters, also providing an explanation for the onset of anoxic conditions in the oceans.

Be it as it may, the End-Permian mass extinction is bound to fuel passionate discussions for years to come.

The end of the Triassic

Even more so than in previous crises, the Earth took a long time to recover from the Permian–Triassic mass extinction, perhaps as long as five million years if one is to judge from the sterile black shales that accumulated for many meters above the boundary. Most plant and animal groups had to build up anew from a limited pool of surviving species. But in due time, life flourished again in the oceans and on dry land, branching out to form new families of mollusks and fish, insects and plants, amphibians and reptiles, along with whole new classes of land-dwelling vertebrates, including early mammals and dinosaurs.

In the midst of this Triassic recovery, a noticeable mass extinction took place in the oceans: the Carnian event, 225 million years ago. Over 40 % of marine species went extinct at the time, but not enough for the Carnian to be classified as one of the great mass extinctions. However, one can't fail to note that a sizeable impact crater in Russia – the 80-km-wide Puchezh-Katunki astrobleme – happens to be 220 ± 10 million years old.

The fourth of the great mass extinctions took place soon thereafter, marking the close of the Triassic and the advent of the Jurassic period, 205 million years ago. This TR-J crisis – also known as the Norian event – saw the extinction of 60 to 70 % of marine species (only slightly less than the later K-T crisis), with an apparent mass killing event, underscored by a carbon isotope anomaly.

Up the marine food chain, disaster struck coral reefs, ammonoids, gastropods and bivalves. In northwestern Europe, in particular, over 90 % of mollusk species went extinct. Among the larger animals, the great vertebrate predators that inhabited the oceans were also affected, including the dolphin-like ichthyosaurs, and especially the placodonts, which were fully exterminated.

On land as well – although there are fewer fossil beds to assess the damage – there is an abrupt turnover of pollen species and a layer of fern spores indicative of an ecological catastrophe. There also appears to be a marked extinction of vertebrates (especially in North America), but the scarcity of End-Triassic fossils makes the census particularly difficult.

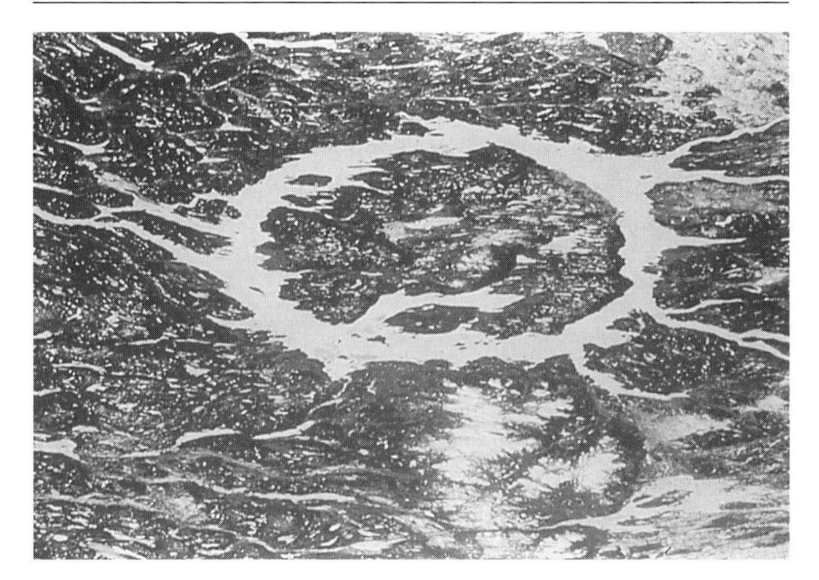

Figure 7.2. The Manicouagan astrobleme in Quebec is 100 kilometers in diameter. The central uplift of crustal rock is surrounded by a moat occupied by a lake. Although this major impact is presently dated at 214 million years, it is believed by some to be slightly younger, and account for the 205 million-year-old End-Triassic mass extinction. (*Photograph: NASA/LPI.*)

This rarity of sedimentary sections spanning the TR-J boundary makes it difficult to identify the cause of the great mass extinction. There is isotopic evidence of a short and massive killing event at the boundary, as we just noted. Iridium anomalies, on the other hand, are so far lacking. But in 1992, David Bice and colleagues from Carleton College and Syracuse University found abundant shocked quartz at the boundary in Italy. In a small gorge near the village of Corfino in northern Tuscany, the scientists identified three distinct layers of shocked minerals, one at the boundary itself and the two others approximately one and two meters below the boundary, possibly indicating three continental impacts within less than 200 000 years.

There happens to be a conspicuous impact crater of roughly the right age: the 100-km-diameter Manicouagan astrobleme in Quebec, dated at 214 ± 2 million years (see figure 7.2). There is no doubt that such an impact must have devastated North America, where vertebrate extinctions at the TR-J boundary are indeed significant. But there is a catch: the Manicouagan impact appears to precede the mass extinction by about nine million years (214 ±2 m.y. versus 205 m.y. for the extinction). If Manicouagan did strike 214 million years ago, why is there no evidence of a biotic crash at that time? The other possibility is that the age determination of the crater or the TR-J boundary is in error, and that future measurements will show them to match more closely.

Indeed, the evidence that the End-Triassic extinction was caused by impact has been bolstered by a joint study in 1998 by the University of New Brunswick (Canada), the Open University at Milton Keynes in the U.K. and the University of Chicago. John Spray, Simon Kelley and David Rowley have shown that the Manicouagan impact structure is aligned with two other craters of similar age, that were at the same latitude at the time of impact (22.8 °N): Saint Martin in Manitoba (40 km in diameter) and Rochechouart in France (23 km). Their conclusion is that a comet broke up at the end of the Triassic, spraying the Earth with several projectiles, as our planet rotated under the line of fire. As we shall see in the next chapter, such a scenario was played out under our very eyes, when comet Shoemaker–Levy 9 bombarded planet Jupiter in July of 1994.

Finally, one must note that the End-Triassic crisis was the last to hit a unified continental block. At the onset of the Jurassic period, the Pangaea supercontinent began to break up: rifting and eruption of trap basalts led to the opening of the North Atlantic Ocean and to the drifting apart of North America and Africa.

The end of the Jurassic

The Jurassic period saw the spectacular rise of the dinosaurs on land. By the Late Jurassic they had achieved dominance over the terrestrial ecosystem, with the famous long necked *Brachiosaurus* and *Diplodocus* browsing the tree canopies, while stegosaurs and primitive iguanodonts roamed the underbrush. Mammals were evolving as well, keeping a low profile in the shadow of their distant cousins. Pterosaurs and other flying reptiles were also diversifying, as were crocodiles, turtles and lizards.

The prosperous, climate-stable Jurassic period came to an end 144 million years ago with a minor mass extinction – the Tithonian event – during which a third of all marine species were exterminated. On land as well there seems to have been a sharp decline in dinosaurs and other four-legged vertebrates at the Jurassic–Cretaceous transition, along with a change in the balance of fauna (among the dinosaurs, sauropods and stegosaurs lost their prominent role to the ornithopods).

The evidence for cosmic intervention at the Jurassic–Cretaceous boundary is manifest. An early report of a strong iridium anomaly in Siberia was for long the only trace of impact, but there is now much more. One already knew of the existence of a small impact crater in Australia, the 22-km wide Gosses Bluff structure (see figures 4.4 and 4.5), dated at 142 ± 0.5 million years, which is dangerously close to the age of the Jurassic–Cretaceous boundary (144.2 ± 2.6 million years).

Even more convincing is the Mjølnir underwater astrobleme, identified in 1993 in the Barents Sea, off the northern coast of Scandinavia. Recent examination of oil exploration drill holes in the area shows that the crater's ejecta – containing iridium and shocked quartz – occurs at the Jurassic–Cretaceous boundary, as indicated by the turnover of ammonite and bivalve species. Mjølnir crater is 40 km in diameter (see figure 7.3).

Besides Gosses Bluff in Australia and Mjølnir in the Barents Sea, one must add the recent discovery of yet another astrobleme in South Africa. Christian Koeberl of the University of Vienna and co-investigators from Johannesburg and Australia analyzed cores from a buried circular structure near Morokweng. Measuring at least 70 km in width, according to magnetic field data, the Morokweng structure contains impact melt rich in meteoritic elements, and zircons from the cores yield ages of 146.2 ± 1.5 and 144.7 ± 1.9 million years, indistinguishable from the Jurassic–Cretaceous boundary.

Finally, there is a possible impact crater in north China – the Duolun structure – which is roughly 100 km in diameter and dated

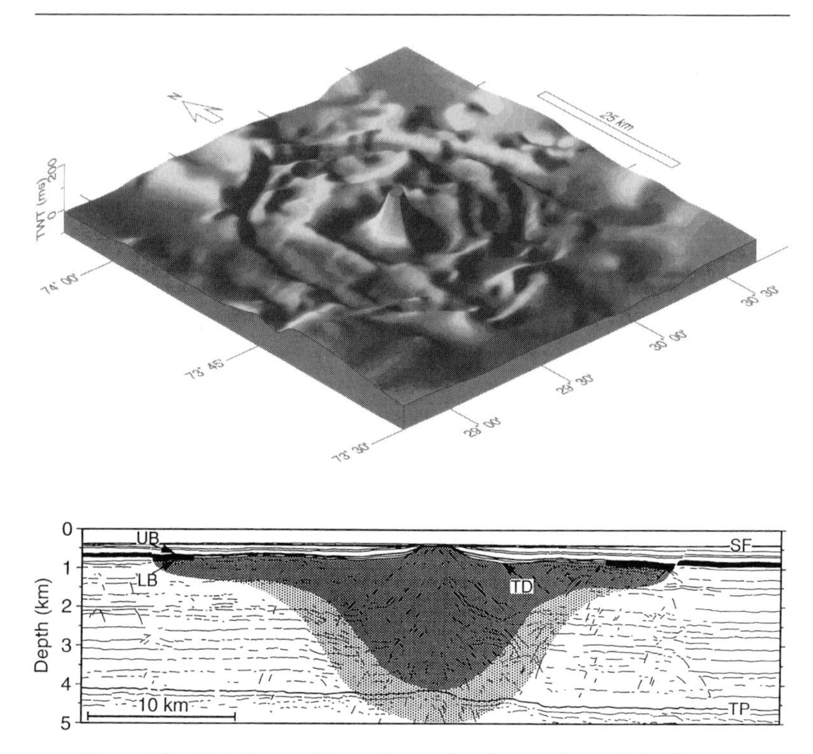

Figure 7.3. Mjølnir astrobleme lies under the sea-floor of the Barents sea. The crater is dated at 144 million years, and appears to be synchronous with the End-Jurassic mass extinction. Above: perspective diagram of the 40-km-diameter crater, derived from seismic reflection profiles (vertical exaggeration is 20×). Below: cross-section of the crater from seismic data. Dark shading indicates greatest disturbance of the sedimentary layers. (*Courtesy of Dr Henning Dypvik. From: Mjølnir structure: an impact crater in the Barents Sea,* Geology, *24, 779–82, reproduced with the permission of the authors and* Geology.)

at about 136 million years. Its impact origin and age need to be confirmed, but even without Duolun, the Mjølnir and Morokweng astroblemes strongly suggest that the Cretaceous period was ushered in by impacts, just as clearly as it was terminated by one at the K-T boundary.

In between the Mjølnir/Morokweng and the Chicxulub impacts, the Cretaceous period itself, from 144 to 65 million years ago, was relatively spared, except for a couple of minor mass extinctions in

Figure 7.4. Small mass extinctions in the fossil record also show evidence of impact. The Cenomanian–Turonian boundary, 91 million years ago, is associated with a field of impact breccia (dark boulders) that crop up on the Atlantic coast of Portugal, near Nazaré. (*Photograph: courtesy of José Fernando Monteiro, Lisbon University.*)

the marine realm, and one regional extinction in North America, 74 million years ago.

The latter is of particular interest since it matches the age of the Manson impact structure in Iowa (see chapter 4). The impact hit the shallow continental sea that covered the North American plains, leaving a 35-km-wide astrobleme and associated tsunami deposits. When the age of the Manson impact was pinned down at 74 million years, paleontologist Dale Russell pointed out that it coincided with a regional turnover of terrestrial and marine species. Many species of crocodiles, plesiosaurs (great swimming reptiles), dinosaurs and mammals were apparently exterminated along the central seaway, and replaced by new populations from outside the devastated area, which came from Asia through the Bering Strait.

This is the first regional extinction convincingly tied to an impact. Less than ten million years later, the great Chicxulub impact provided the showcase example of a global, planetary mass extinction.

Extinctions in the Mammal age

The K-T mayhem, 65 million years ago, was the fifth and last great mass extinction of the fossil record. But the rise of the mammals in the Tertiary era, culminating with the ascent of man, was rocked by several minor extinctions, starting with the Late Eocene events around 35.5 million years ago.

This was evidently a drawn out, multiple event with pulses of extinction rippling through the marine realm: no less than four plankton crashes occurred in less than a million years, together with extinctions of gastropods and bivalves. On dry land, there was a significant turnover of the mammalian fauna in North America and Asia; likewise in Europe one notes the appearance of new species of carnivores and rodents in the aftermath of the crisis, as well as a marked change amongst reptiles and amphibians, to the point where French paleontologists nicknamed this transition *la grande coupure* ('the great cut').

As it turns out, the Late Eocene is teeming with evidence of multiple impacts, clustered between 36 and 35 million years ago. Alessandro Montanari reported iridium anomalies (up to 4 ppb) at several localities, sometimes doubled up into two layers. Moreover, several layers of tektites and microtektites are known from the Late Eocene, including a strewn field of tektites off the U.S. East Coast, extending from New Jersey to the Caribbean, and associated with some shocked quartz and other impact-related minerals (coesite and lechatelierite). There is also an apparently slightly older layer of microspherules stretching from the West Atlantic to the equatorial Pacific and Indian Ocean. To top this off, there is also shocked quartz in Italy, reported by Aron Clymer, which is associated with a spike of iridium and seems to represent the first of several impacts, 35.7 million years ago.

A candidate crater for the Late Eocene crisis has been known for years: the Popigai astrobleme in Russia, 100 kilometers in size and dated at 35 ± 5 million years (a recent measurement seems to narrow down the age to 36 ± 1 million years). The shocked quartz from Italy might very well be Popigai ejecta, since it is mineralogically similar to quartz in the Russian target area. The Popigai impact, then, would have opened the Late Eocene salvo, shaking up the marine realm with a major turnover of bottom-dwelling *foraminifera*.

More of a surprise is the recent discovery of a major impact in Chesapeake Bay, downstream from Washington D.C. After discovering a boulder bed in the area, which they interpreted as an impact-generated tsunami deposit, Wiley Poag and partners used seismic profiling to identify a buried 90 km diameter astrobleme in the bay, combined with a smaller crater, 25 km in size, off the coast of Atlantic City. The Chesapeake astrobleme – the largest in the U.S. – is convincingly tied in to the tektite field of the area. Moreover, the tektite layer is directly associated with an extinction of radiolarians (silica-shelled microorganisms) that close the sequence of biological crises of the Late Eocene, approximately 35.4 million years ago.

The double hit of two 100-km-class craters – Popigai and Chesapeake – apparently left a distinct, albeit modest extinction record at sea and on land. There was apparently no greenhouse effect in this drawn out crisis, but a global cooling, as is apparent from oxygen isotope anomalies and a pulse of ice-rafted sediment in the southern Indian Ocean, marking the onset of glaciation in Antarctica.

Altogether, judging from the micrometeorites and ablation spinels dispersed in the sediment (spanning anywhere from 10 000 to a million years), Eric Robin estimates that the cosmic salvo spread ten times less meteoritic material around the globe than the Chicxulub impact, and supports the theory of a comet shower, previously suggested by Piet Hut and others.

After the Late Eocene crisis, the Earth was relatively spared by impacts up to the present time. There was a minor mass extinction in the middle of the Miocene period, around 14 million years ago, marked by a modest iridium anomaly and glass spherules that might be meteoritic. The famous, well preserved Ries crater in Germany, 23 km in size, is dated at 15 ± 1 million years.

Closest to us, Late Pliocene sediments, just over two million years old, bear a significant iridium anomaly (5 ppb), with other rare metals in cosmic proportions as noted by Frank Kyte, and a layer of microspherules interpreted to be impact ejecta.[9] Frank Kyte and co-workers have identified the impact site – named Eltanin – in the southeastern Pacific, where it takes the form of a chaotic sedimentary layer of breccia, overlain by layered sand and clay, a sequence strikingly similar to the Chicxulub tsunami deposits in Mexico. There is no crater, because the 1 to 4 km asteroid was not big enough to blast through the deep ocean and the sediment, down to the crust. Had it occurred on land, the crater would have measured 30 to 40 km in diameter. This is not big enough to cause a mass extinction (none is observed) but this impact does seem to coincide with a strong cooling event, 2.15 million years ago, near the beginning of the Pleistocene ice age.

Making sense of the record

What conclusions can then be drawn from this global review of mass extinctions? Do they correlate with impacts in a convincing way?

Out of 25 mass extinctions tentatively identified in the fossil record,[10] six are associated with significant evidence of impact (large iridium anomalies, shocked quartz and/or microtektites). Chronologically, they are the Late Devonian (F-F); End-Triassic (TR-J); End-Cretaceous (K-T); Late Eocene; Mid-Miocene; and Late Pliocene events. The first three are major mass extinctions; the last three are minor ones.

Seven other mass extinctions are associated with lesser iridium anomalies, many of which could also have a cosmic origin: the dawn of the Cambrian period; the End-Ordovician; End-Devonian (D-C); End-Permian (P-TR); Callovian; End-Jurassic; and Cenomanian events.

That between a quarter and a half of all mass extinctions in the

[9] Chinese scientist H. Peng also discovered microtektites in North Pacific cores, synchronous with a plankton extinction. Similarly, microtektites have been discovered in Late Pliocene soil in the Chinese province of Shanxi.

[10] The 25 mass extinctions identified by Jack Sepkoski in the marine fossil record, including the Precambrian–Cambrian transition.

Figure 7.5. Impacts can shower the Earth in salvos, where a broken-up comet or a cluster of asteroids intersects the Earth's trajectory. Such cosmic salvos might account for the evidence of multiple impacts found at the End-Devonian, End-Jurassic and Mid-Miocene boundaries. (*Photograph: courtesy of William K. Hartmann (© William K. Hartmann).*)

fossil record appear to be connected with impacts is remarkable, especially since research in the field has only just begun. More correlations might emerge in the future. One can then suggest that impacts are the principal mechanism of biological crises on this planet (which doesn't preclude other mechanisms from occasionally coming into play). However, such a postulate is convincing only if it works the other way around as well: are all large astroblemes on Earth associated with mass extinctions? Are there any notable exceptions that weaken the correlation?

If we look at impact structures larger than 80 km in diameter (our theoretical threshold for a global blackout of sunlight), all but one coincide with a mass extinction. Puchezh-Katunki crater (80 km) is contemporaneous with the small Carnian mass extinction; Manicouagan (100 km) might be the exception, not squaring with any mass extinction (but it is 'dangerously' close to the Tr-J boundary); Chicxulub (180 km) is clearly associated with the K-T catastrophe;

and both Popigai (100 km) and Chesapeake craters (90 km) are synchronous with the Late Eocene mass extinction.[11]

Manicouagan is the only awkward crater, since it is apparently seven million years older than the End-Triassic mass extinction. If nothing is wrong with the dating, here is evidence that a large impact, slightly above the theoretical 'threshold of harm', did not trigger a mass extinction. Why so?

One can only speculate, but three reasons come to mind, if the dating is not in error.

One is that there was no target effect during impact, and few volatiles like carbon or sulfur were spewed into the atmosphere. However, one finds carbon-rich limestone blocks in the Manicouagan melt sheet, which hints that the impact did kick up a substantial amount of carbon dioxide into the atmosphere.

Another possibility is that Manicouagan was a near vertical rather than oblique hit, with only minor coupling of the impact energy with the atmosphere, and little environmental damage.

A third possibility is 'immunization' of the ecosystem. An 80-km-sized impact occurred a mere six million years before Manicouagan, probably connected with the massacre of 40 % of the species living at the time (Puchezh-Katunki and the Carnian extinction). Survivors were obviously those species most resistant to impact disturbances, so that the next hit (Manicouagan) might have added few new victims to the list, and did not result in a noticeable mass extinction.

But then, why didn't this immunization concept work one more time, when a new large impact, the one testified by the iridium and the shocked quartz, *did* trigger a mass extinction seven million years later (the End-Triassic). Clearly, the inconsistencies would vanish if Manicouagan crater was younger than its presently estimated 214 million years and squared with this End-Triassic extinction. Wishful thinking or logical solution?

[11] The 100 km Duolun structure in China has not yet been confirmed as an astrobleme, nor dated with sufficient accuracy to be considered in this list. But it is tentatively associated with the End-Jurassic mass extinction, as is Morokweng astrobleme in South Africa, which could be much larger than 70 km.

The volcanic alternative

With the possible exception of Manicouagan, most large impacts show a convincing correlation with the mass extinctions that punctuate the history of life on Earth.

There is a competing unifying model to explain mass extinctions: catastrophic volcanism, driving climate upheavals. In chapter 3, we saw how the Deccan Traps of India were offered as an alternative explanation for the K-T crisis. The synchronicity of the Siberian Traps with the End-Permian mass extinction has further strengthened the case of volcanists, who have also signaled a number of other trap eruptions to be synchronous with mass extinctions. Compiling the data, Michael Rampino and Richard Stothers have confirmed what appears to be a significant correlation. French geophysicist Vincent Courtillot has done so as well, entitling one of his articles 'Mass extinctions: seven traps and one impact'.

The correlation between mass extinctions and trap eruptions is a mixed bag. If we examine them in chronological order, the Siberian Traps (250 million years ago) do coincide with the great End-Permian mass extinction. But, second in line, the Eastern North American Traps (202 m.y. ago) follow rather than strictly coincide with the End-Triassic mass extinction (205 m.y. ago), although the time difference could turn out to be less than a million years.[12]

Then come the Karroo Traps of South Africa (around 190 m.y. ago), which can argumentatively be tied in with the small Pliensbachian extinction (191 m.y. ago). On the other hand, the subsequent Ferrar Traps of Antarctica (around 180 m.y. ago) do not coincide with any mass extinction. Neither do the great Parana-Etendeka Traps of South America and South Africa (132 m.y. ago). Closer to us in time, the mixed bag continues: the small Rajmahal Traps of India (around 115 m.y. ago) might correlate with the small Aptian mass extinction (112 m.y. ago). The Deccan Traps of India did start a couple of million years *before* the End-Cretaceous mass extinction (65 m.y. ago), and overlapped the boundary. The Brito-

[12] In the field, the first lava flows of the New Jersey Traps stand 40 centimeters *above* the fern spike and the mass extinction boundary, thus 400 000 years *later*, if we assume a sedimentation rate of one millimeter per millennium.

Arctic Traps (59 m.y. ago) are also close to the Paleocene–Eocene boundary (a small mass extinction). But the correlation is lost for the Ethiopian Traps, now dated at 30 million years – five million years *younger* than the Late Eocene mass extinction (35 m.y. ago), and for the Columbia River traps (17 m.y. ago), which are three million years *older* than the minor Mid-Miocene extinction (14 m.y. ago).

In summary, about half (four) of the traps appear unrelated to mass extinctions (Antarctic; Parana-Etendeka; Ethiopian; and Columbia River traps). One trap is synchronous with a mass extinction (Siberian Traps and P-TR crisis); and another starts at least two million years before, and overlaps a mass extinction (Deccan Traps and K-T boundary). Two other traps might coincide with minor extinctions (the Karroo with the Pliensbachian, and the Rajmahal with the Aptian extinction). Lastly, there is one case where a trap closely *follows* a mass extinction (the Eastern North American Traps coming after the TR-J crisis).

The correlation that emerges in the five latter cases can be interpreted in different ways. One scenario is that traps and mass extinctions occur fortuitously on the same timescale (20 to 30 million years between events) and that they happen to be in phase, without necessarily being related in a causal way. This can be said of the Deccan Traps, which are one of the three larger trap eruptions on record, but did not cause any noticeable mass extinctions at the peak of their activity, around 67 million years ago (when dinosaurs, as we saw in chapter 3, were thriving in India). The K-T mass extinction occurred on its own, in the waning stages of the volcanic period, and is due to an impact.

This lack of harmful effects of major traps is confirmed by the Parana-Etendeka Traps, which are just as voluminous, fast paced and even more siliceous and explosive than the Deccan Traps, and have left no trace of a mass extinction in the fossil record.

On the other hand, the eruption of the great Siberian Traps does match the great End-Permian mass extinction, and the Eastern North American Traps erupt very soon after an extinction event. Here, it can be argued that these traps were triggered by impact, and thus were the effect, and not the cause, of a mass extinction event.

Looking for cycles

That mass extinctions, impacts and volcanic traps appear to follow cycles is hardly surprising. Nature is fond of cycles, which go beyond the simple realm of hazard and are driven by very real mechanisms. Identifying a cycle can thus lead to the discovery of the driving mechanism itself.[13]

It is with this idea in mind that A.G. Fischer and Michael Arthur in 1977, and David Raup and Jack Sepkoski in 1984 analyzed the record of mass extinctions in search of a periodicity. They came to the initial conclusion that biological crises occur on average every 26 million years. Further analyses by a number of scientists, using sophisticated mathematics, suggest a range of periods between 26 and 31 million years.[14] Recently, Michael Rampino and Bruce Haggerty have reprocessed the data and settled on a 26.5 to 28 million year periodicity.

The record of impact craters has also been searched for potential cycles. Unfortunately, the cratering record is notoriously incomplete and few of the discovered craters are precisely dated. Despite these shortcomings, astronomer Piet Hut suggests that the number of astroblemes appears to surge at 95, 65 and 35 million years ago and in recent times, showing a periodicity of roughly 30 million years.

Are these apparent periodicities connected with a fundamental driving mechanism? On the one hand, one can interpret them as simply reflecting the probability of occurrence of events of a given magnitude. For instance, given the fact that the collision frequency of asteroids and comets is inversely proportional to their size, a 26 to 30 million year period might merely represent the 'waiting time' between strikes by impactors a few kilometers in size, blasting astroblemes large enough (80 to 100 km?) to perpetrate mass extinctions.

[13] In the case of trap eruptions, for instance, Vincent Courtillot believes their periodic occurrence reflects a cycle of thermal instabilities deep within the Earth at the mantle-core boundary: this cycle would trigger both the ascent of magma plumes and inversions of the Earth's magnetic field.

[14] A.G. Fischer leans toward a longer 74.5 million year periodicity in the occurrence of mass extinctions. The End-Permian crisis lies outside this cycle, which might reflect an extraordinary origin.

Figure 7.6. The apparent periodicity of mass extinctions – every 30 million years or so – brought researchers to propose a cyclic driving mechanism. One such mechanism could be the periodic incursion of comets into the inner Solar System, where they would impact the planets. (*Photograph: by the author.*)

On the other hand, some astronomers believe that the observed periodicity is the doing of a specific, tangible mechanism that focuses asteroids or comets into 'herds', periodically crossing the inner Solar System and threatening the Earth.

This has led to the suggestion that a small companion star orbits our Sun, far enough (three light years) and faint enough (a red dwarf) to have evaded detection. Circling its eccentric orbit in 26 million years or so, it would periodically swing in close to the comet cocoon that marks the outer limits of our planetary system, and dislodge some of the icy bodies to send them hurtling towards the inner planets.

The theory of the companion star was popular in the mid-eighties, to the point where the hypothetical stellar body was named Nemesis, and efforts were made to locate it with telescopes. However, not only was the search unsuccessful but computer simulations have shown that the orbit of such a companion could not be stable (perturbed as it would be by other stars, as our system travels through the Galaxy). Likewise, the possibility of a perturbing tenth planet (rather than a star) has failed to convince astronomers.

A more plausible mechanism for periodicity was proposed by Rampino and Stothers in 1984:[15] they called attention to the motion of the Solar System through the Milky Way Galaxy, which brings it to periodically cross the galactic plane where cosmic matter (gas and dust) is denser than average. The half-period of this oscillation (the time between galactic crossings) is estimated to be anywhere from 26 to 36 million years, depending on galactic models.

According to recent calculations by Matese *et al.*, the resulting 'galactic tides' would increase the flux of comets toward the inner Solar System (and thus the probability of collisions) by a factor of five over an interval of several million years. Named the *Shiva hypothesis* by Rampino and Haggerty (after the Hindu god of destruction and renewal), the model of periodic comet salvos is particularly disquieting in that the next peak in the 30 million year cycle is . . . now.

This leads us in turn to examine the frequency and probability of impacts today, within the timeframe of human civilization, and what we can do to deter or avoid them.

[15] The idea was adopted later that year by astronomers Clube and Napier.

8
The impact hazard

After the discovery of Chicxulub crater and its recognition as the cause of the K-T mass extinction, after the realization that other mass extinctions were probably also due to impacts, the next logical step is to consider how much of a threat impacts are to us, here and now.

There is some belief, even in scientific circles, that there is no serious risk of a devastating impact on the timescale of human civilization, but this is an erroneous belief, due mostly to lack of information. In fact, a look at the historical record shows that impacts have already struck home in recent times.

Hell in Siberia

We need not go back many years to recall a devastating impact that not only claimed lives, but flattened and roasted thousands of acres of forest land with an explosive force estimated at 15 megatons of TNT – one thousand times the energy of the Hiroshima bomb.

This impact took place on June 30, 1908, a little after sunrise, in the remote taiga of Siberia near the river Tunguska (65 °N, 102 °E). Fortunately the area at ground zero was unsettled. The closest outpost was at Varanova, a trading station 70 km away, and the closest village – Nizhne-Karelinsk – was farther yet, along the line of the Trans-Siberian Railroad. Even at such distances, the blast was horrendous. At Varanova, glass windows were blown out by the air wave and bystanders knocked to the ground. The thunderous rumble was heard 600 km away, over an area nearly the size of France. Barographs recorded the jump in atmospheric pressure all around the world, as the blast wave circled the globe a full three times . . .

Eyewitnesses of the Tunguska event – as it came to be known – gave breathless accounts. Workers on the Trans-Siberian Railroad spoke of a brilliant fireball over the northern horizon, seconds before the explosion, followed by a blinding flash and a column of fire illuminating the sky. At Nizhne-Karelinsk, the radiance was so strong that villagers had to divert their eyes. Panic-stricken men ran into the streets and women wailed, fearing that the end of the world had come.

Nearly twenty years were to pass before the first scientists got a chance to visit the remote area of the impact.[1] They found out that there *had* been victims. Interviews with nomadic herdsmen established that about twenty of them were within 50 km of ground zero at the time of the impact. All were toppled and injured by the blast, experiencing severe deafening and shock trauma. One of the elders was tossed ten meters against a tree trunk and died of his injuries; another died of sheer shock. As for the herds they were tending, an estimated one thousand reindeer were killed and charred by the fiery blast, along with a number of herd dogs.

Close to ground zero, explorers like Leonid Kulik recorded the utter devastation that had befallen the forest. Strangely, there was no crater to be found. Instead, the trees lay flat, toppled by the blast up to 30 km from ground zero. Deep into the target area, inside a circle 15 km in radius, the tree trunks were charred. Ground zero itself was conspicuous in that charred tree trunks were left standing, as if the blast had occurred some distance above ground level, blowing downward in the center and only stripping off the branches, whereas farther out the trees were blown down by the radial component of the blast. The devastated area covered 2000 km^2 – the size of a large city – with a slight elongation in a northwesterly direction, as if the impactor had flown in obliquely from the southeast.

[1] Not only was the boggy taiga remote and difficult to access, but social unrest – leading up to the Russian Revolution – delayed all scouting expeditions for over a decade. It was not until 1924 that geologists were able to interview the herdsmen who had been closest to the impact, and 1927 before a scientist from Leningrad, Leonid Kulik, was able to reach the devastated area.

The absence of a crater can be explained by the fact that the projectile apparently disintegrated in mid-air, some 8000 to 6000 meters above the ground. The small projectile had low density and strength, as would a fragment of a comet or a stony asteroid. It is believed to have measured roughly 50 to 60 meters in diameter, with a kinetic energy of 15 megatons of TNT. Had the projectile hit the ground, the resulting crater would have measured close to a kilometer across, almost the size of Meteor Crater in Arizona.

It takes little effort to imagine the consequences that such an impact would have had on a populated area. Isaac Asimov pointed out that if the impactor had intercepted the Earth a mere five hours later, the time for the planet to spin some 70° in longitude, the city of St Petersburg (Leningrad) could have been wiped off the map, along with its hundreds of thousands of citizens. Today, a blast like Tunguska's over a major city would claim millions of victims.

Because it struck a remote area, well before the advent of mass communication, the Tunguska event went relatively unnoticed in the scientific arena and in the popular consciousness. But its significance kept growing over the years as more scientists caught on to its cosmic origin and power of destruction.[2] The Siberian blast has become a textbook example of the destructive mechanisms at play in a cosmic collision,[3] and influenced the 'nuclear winter' studies of the 1980s, which underscored the perverse climatic effects that even a small nuclear exchange (or impact) would force upon the biosphere.

The Tunguska event of June 30, 1908 stands as startling proof that impacts are not confined to the distant past and are a very real threat to human life today. Our knowledge of the population of

[2] The Tunguska event was one of the pieces of evidence that led paleontologist M.W. De Laubenfels to propose, as early as 1956, that mass extinctions like the K-T crisis were due to impacts (see chapter 2).

[3] Indeed, the 'small' 15-megaton blast at Tunguska provides a scaled-down version of the mechanisms discussed in chapter 6 in relation to the Chicxulub impact: blast air wave and hurricane-force winds; fireball and ignition of fires on the ground; chemical reactions in the atmosphere and aerosols disturbing the climate. Regarding the latter, it has been calculated that up to 50% of the ozone in the northern hemisphere was momentarily destroyed by the Tunguska impact, as nitrous oxides formed in the shock wave. According to this model, the ozone level recovered in a matter of days. As for the aerosols in the upper atmosphere, they cast an eerie glow for several nights over northern Europe, and led to an atmospheric cooling that reached about one degree by the end of 1908.

small objects – asteroidal and cometary – crossing the Earth's orbit tells us that 15-megaton impacts of the Tunguska class, capable of wiping out entire cities, occur *every century* on average. Impacts of lesser magnitude are even more frequent.

The historical record

The Tunguska event of 1908 is the most spectacular example of a lethal impact in recent times, claiming two victims (herdsman Vasily and hunter Lyuburman), and massacring over a thousand reindeer. But impact casualties are not limited to the Siberian event alone. If one looks up news reports and other printed records of the last few centuries, the number and variety of occurrences of meteorite hits are astonishing.

By combing over the historical records, scholars have compiled many accounts of death by impact. Notable are Kevin Yan, Paul Weissman and Don Yeomans at the Jet Propulsion Laboratory who have searched ancient Chinese records, and John S. Lewis, Codirector of the Space Engineering Research Center in Tucson, who has reported his findings in his book *Rain of Iron and Ice*.[4]

In the twentieth century alone, apart from Tunguska, Lewis lists a fatal meteorite crash in Zvezvan, Yugoslavia, killing one person at a wedding party in 1929; and the death of an entire family in 1907, crushed by a meteorite fall in the Weng-li district of China.

In the course of the century, it is also worth noting that five cars have been struck by meteorites without loss of life: in 1938 (Illinois), 1950 (Missouri), 1965 (England), 1977 (Kentucky) and 1992 (New York). Because of the veneration surrounding automobiles – and their market value as collectors' items – these car hits have been widely publicized, so much so that one might wonder if they attract meteorites in some mysterious way! However, a look at John Lewis's compilation shows that buildings register many more hits. A stag-

[4] John S. Lewis, *Rain of Iron and Ice, the very real threat of comet and asteroid bombardment*, Helix Books, Addison-Wesley Publishing Company, 1996, 236pp.

gering total of over 50 strikes on houses, sheds, barns, farmhouses and other buildings are reported in the twentieth century alone.

To round off the list, there are two cases of planes sweeping through – and surviving – meteor showers;[5] and two cases of boats being hit, in 1936 and 1938, the first being set ablaze by the strike. At least two forest fires are also attributed to impacts, in Mexico (1910) and Massachusetts (1921). Finally, among the most humorous cosmic strikes, one mailbox was demolished by a meteorite in Georgia (1984); a meteorite flew into a room through an open window in Japan (1949); and the dome of an amateur telescope was pierced by a pair of meteorites in Washington State (1955), setting astronomy books on fire.

If we return to the more somber study of casualties, the apparent toll in the twentieth century reaches about half a dozen victims. The nineteenth century boasts a similar record, with one man killed and one woman injured in India in 1825; a child killed in China in 1874; and a French farmer killed in 1879.[6]

Many more of these small – but potentially deadly – meteorite hits certainly took place throughout the centuries and were not recorded, for lack of written records or simply lack of witnesses. Perhaps the number of sporadic casualties from meteorites reaches several dozen victims per century.

This estimate describes only the 'background' death rate, without calling into play exceptional mass killings that might occur at less frequent intervals. Indeed, meteorite falls occasionally include boulder-sized objects that split into salvos of fist-sized 'shrapnel', riddling the land below with stone and iron. Such meteorite showers have been reported several times in the past, notably at L'Aigle, France (1803); Holbrook in Arizona (1912); Sikhote-Alin in Siberia (1947); and Allende Pueblo in Mexico (1969). These showers struck open fields and claimed no victims, but one fatal shower apparently did take place in the Sanshi province of China in the winter of 1490, claiming a reported 10 000 lives. If this massacre is truly due

[5] Both in 1934, for some reason!

[6] Among other sporadic deaths reported in previous centuries, a French farmer was also killed in 1790; two sailors were killed on the deck of a ship in the 1650s; and two Italian monks were reportedly hit and killed by meteorites in the sixteenth and seventeenth centuries.

to a meteorite salvo, here is proof that unfortunate targeting of a meteorite shower can have dramatic consequences.

Larger and rarer impacts of the Tunguska class, that can take out an entire city, strike every century on average, but they are expected to claim victims only in a minority of cases. Indeed, four out of five impacts occur at sea or in the unpopulated polar zones. Tunguska itself, the only case on record, occurred in an unpopulated area of Siberia.

It is also a fact that blasts of Tunguska magnitude usually occur above ground and leave little evidence for geologists to later discover. Fallen trees, for example, rot away in a matter of decades, so that there is little hope of discovering old scars of such 'airburst' impacts.

Despite these difficulties, Duncan Steel and Peter Snow signal the possible site of a Tunguska-like impact in the South Island of New Zealand. Fallen trees point radially out from a hypothetical ground zero – an elongated depression 900 by 600 meters in size – over a radius of 80 kilometers. The impact apparently occurred 800 years ago, at a time when the South Island of New Zealand experienced widespread fires and the local moa bird went extinct. The mythology, poetry and song of the Maori people all tell of the falling of the skies in the South Island, accompanied by raging winds and devastating fires.

The mythical record

So be it for small impacts on a timescale of a millennium or so. We next ask ourselves what to expect, in terms of impact size, on a longer timescale of 10 000 years – the duration of human civilization. Are there any signs of large impacts in our myths and legends?

Every 10 000 years, probabilities call for the impact of one or two 500-meter-diameter projectiles, which would blast craters five kilometers across or wider if they hit land, or send momentous tsunami racing across the ocean if they struck water.

On land, such an impact would flatten and char up to 150 000 km^2, an area the size of Belgium. At sea, it would propel an ocean wave hundreds of meters in height, a swell that would still measure tens of meters in amplitude when it reached a continental shore, so that – compounded with the tenfold surge in height generated by friction of the wave on the shallowing sea bottom – the tsunami would tower hundreds of meters over the shoreline, and rush hundreds of kilometers inland. Fireball, ejecta and noxious gases all would add to the coastal devastation.

Since a blast of this magnitude is expected to occur every 5000 to 10 000 years, mankind might already have experienced a large disaster since the days when it emerged from the caves of the ice age to establish its agricultural and social foothold. We can look for clues of such a catastrophe in the myths and legends handed down from the early days of civilization. Some myths might well be rooted in fact, and there is a notable wealth of impact and flood stories that repeatedly crop up in about every culture and religion on Earth.

In the Bible's Book of Revelation, for example – which was written in the first century AD but recounted stories that were probably passed down many generations – there are vivid accounts of what appear to have been cosmic impacts. In the ominous opening of the sixth seal of Revelations, St John writes that 'the sun became black as sackcloth of hair, and the moon became as blood; and the stars of heaven fell unto the earth . . .'

As for the opening of the seventh seal, '. . . there followed hail and fire mingled with blood . . . and the third part of trees was burnt up; and all green grass was burnt up . . .'

'And the second angel sounded, and as it were, a great mountain burning with fire was cast into the sea . . .'

'And the third angel sounded, and there fell a great star from heaven, burning as it were a lamp . . .'

The fourth angel goes on to sound obscurity over the Earth; and the fifth angel ushers in yet another impact: 'I saw a star fall from heaven upon the earth . . . and there arose a smoke out of the pit, as the smoke of a great furnace; and the sun and the air were darkened.'

A number of scholars have researched biblical and other early records of human history in search of such accounts. At the University of Vienna, Edith Kristan-Tollmann and Alexander Tollmann undertook a monumental compilation of world mythology and religious writings, and concluded that many stories pointed to a common, global impact experience which they tentatively dated at 7500 BC, citing the (disputed) age of tektites found in Australia. Interpreting literally the number seven (which is elevated to mystical importance in several legends), the authors postulate that the impact consisted of a salvo of seven cometary fragments around the world – a plausible scenario in light of our recent studies of cometary break-up.

Be it as it may, there must have been devastating impacts by large chunks of rock and ice in the timeframe of human civilization, each with the power to shatter or completely destroy an entire culture. Besides the impacts postulated by the Tollmanns around 7500 BC, Duncan Steel suggests that the period around 3500–3000 BC might have experienced a wave of sizeable impacts (Tunguska-class or larger), and ascribes the rise of astronomy-based monuments in world culture (such as Stonehenge and the pyramids) to a catastrophe-spun urge to survey and predict meteor and cometary showers.

To back up this speculation, there is some evidence of sizeable impacts in the last 10 000 years. Most conspicuous is the oblique impact site of Rio Cuarto in Argentina, a swath of young, elongated craters (1 km wide and up to 4 km long) that extend in a long chain across the pampas. Planetary scientist Peter Schultz estimates that the total energy released in the Argentinian hit was on the order of 1000 megatons, with a 350-megaton blast for the largest crater in the chain.[7]

The hominid record

The fact that traumatic impacts of 500-meter objects occur on timescales of 10 000 years – the lifetime of our civilization – is a sobering and unavoidable conclusion from the study of asteroidal traffic and cratering fluxes in the Earth–Moon system. With a

[7] Another notable string of recent craters is the Henbury Field in Australia. The craters measure 10 to 300 m, and the impact is dated at 2700 BC.

destructive power of hundreds of megatons, these blasts have the potential to topple civilizations.

Lastly, we must consider what greater catastrophes might have occurred on the longer intervals relevant to mankind, on a scale of one or two million years – the timeframe of our emergence as a species from our hominid ancestry.

Logically, on such a longer interval, chances are that even larger impacts occurred: predictions call for a 1-km sized object hitting Earth every 100 000 to 200 000 years. Three hits out of four would be gigantic impacts at sea; the ones on land would leave craters at least 10 km in size. Such impacts would not only directly devastate areas the size of France or Texas, but they would also trigger short-lived climatic collapses of the ecosystem, perhaps dropping worldwide temperatures by several degrees for several months. Today, one such impact would massacre at least a quarter of the human population, mostly because of after-effects like widespread famine.

There are two prominent 10-km-wide astroblemes that testify to such monumental hits during the evolution of man: Zhamanshin in the steppes of Kazakhstan; and Bosumtwi in Ghana (see table 8.1). The Zhamanshin astrobleme is 13.5 km wide and dated at 900 000 years; Bosumtwi is 10.5 km wide and 1.3 million years old (see figure 8.1). The strikes must have spread havoc in Asia and Africa, and one can speculate that their effects were severe on the emerging human race.

Bosumtwi is at the same latitude as, and a mere 2000 miles away from, the Great Rift Valley – the cradle of human evolution. Damage to the climate and perhaps temporary obscurity might have rocked the Rift Valley at a time in evolution that was seeing the rise of *Homo erectus*. Closer to the impact site, the *Australopithecenes* of the Chad plateau would have been affected to an even larger degree.

Later in the story of evolution, the 900 000-year-old Zhamanshin impact in Kazakhstan might also have played some role, at a time that saw the expansion of *Homo erectus* into the Middle East, Europe and Asia, leading to the emergence of *Homo sapiens*.

In the southern hemisphere, another sizeable impact apparently occurred 700 000 years ago, if we are to judge from the extensive field of tektites that extend from China to Australia, but the crater is missing.

Table 8.1 *The most recent impact craters on Earth (over 500 m in diameter). The three youngest ones (Barringer 'Meteor' Crater, Lonar and Amguid) are less than 100 000 years old: they are contemporaneous with* Homo sapiens. *Two astroblemes over 10 km in size – Zhamanshin, Kazakhstan and Bosumtwi, Ghana – are about a million years old and are contemporaneous with* Homo erectus.

Name	Location (Lat, Long)	Diameter (km)	Age (m.y.)
Barringer	Arizona, USA (35 °N, 111 °W)	1.2	0.05
Lonar	India (20 °N, 76 °E)	1.8	0.05
Amguid	Algeria (26 °N, 4 °E)	0.5	<0.1
Wolf Creek	West Australia (19 °S, 128 °E)	0.85	<0.3
Zhamanshin	Kazakhstan (48 °N, 61 °E)	13.5	0.75
Monturaqui	Chile (24 °S, 68 °W)	0.5	1
Bosumtwi	Ghana (7 °N, 1 °W)	10.5	1.03
Tenoumer	Mauritania (23 °N, 10 °W)	1.9	2.5
Talemzane	Algeria (33 °N, 4 °E)	1.75	<3
El'gygytgyn	Russia (67 °N, 172 °E)	18	3.5
New Quebec	Quebec (61 °N, 74 °W)	3.2	1.4
Kara-Kul	Tajikistan (39 °N, 73 °E)	52	<5
Roter Kamm	Namibia (20 °N, 15 °E)	2.5	5
Bigatch	Kazakhstan (48 °N, 82 °E)	7	6
Karla	Russia (58 °N, 48 °E)	12	<10
Shunak	Kazakhstan (43 °N, 73 °E)	3.1	12
Ries	Germany (49 °N, 11 °E)	24	15
Steinheim	Germany (49 °N, 10 °E)	3.4	15

Close calls

Besides the growing census of impact craters on Earth, another sobering lesson comes from the field of astronomy, with the realization that the Earth experiences a number of near-misses with cosmic debris as its circles the Sun.

One shocking warning came on March 23, 1989, with the discovery of a 300-meter-wide rock crossing our orbit at an uncomfortable distance of 650 000 km, less than twice the distance to the

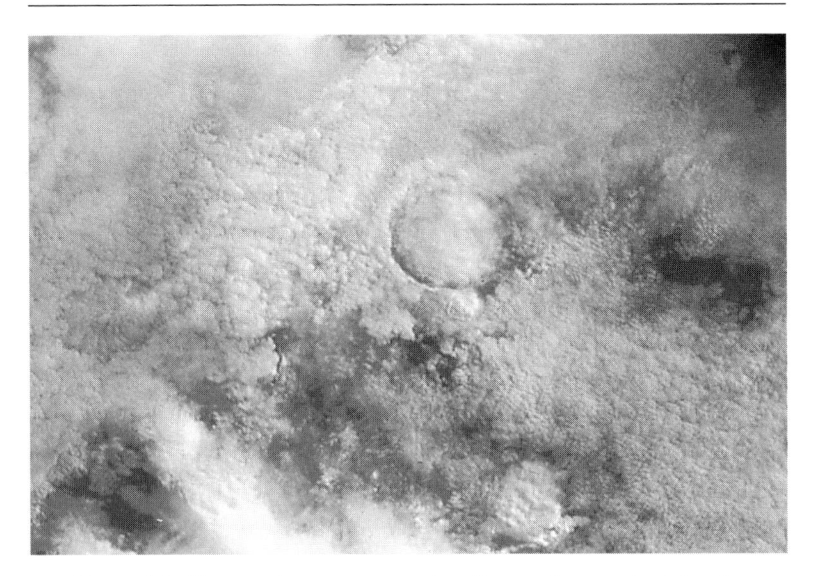

Figure 8.1. Bosumtwi impact crater in Ghana, photographed by the Space Shuttle. The 10-km-diameter crater is filled by a lake and shrouded by clouds. The age of the impact is estimated at 1.03 million years, which makes Bosumtwi one of the most violent impacts since the emergence of our hominid ancestry in Africa. (*Photograph: NASA/LPI.*)

Moon. Had our planet passed six hours ahead of schedule on its swift course around the Sun, it would have experienced on that day an impact of over 1000 megatons, blasting a crater some three to four kilometers wide somewhere on the globe.

After sailing past us, this lethal chunk of rock, numbered 1989FC and baptized Asclepius, will be back on an Earth-crossing trajectory in 2012, but that time around it will pass at a respectable distance.

The 1989 alert has been followed by many others, as improving instrumentation and dedicated searches turn out a growing record of frequent near misses by boulders tens of meters in size. One noteworthy occurrence was the pass of a small cometary 'chunk' near the Earth in March of 1992, observed by astronomers in Chile. The projectile was estimated to have measured 100 to 200 m in diameter and to have missed the planet by a mere 20 000 km. This is a 'minute' miss in terms of orbital schedules, which spared us the horror of a 15 000-megaton hit somewhere on the planet.

Shoemaker–Levy 9

The reality of cosmic collisions was underscored even more graphically in July of 1994 with the greatly publicized collision of comet Shoemaker–Levy 9 with planet Jupiter.

Comet Shoemaker–Levy 9 is named after its three discoverers – Carolyn Shoemaker, Gene Shoemaker and David Levy – who detected the fuzzy object at the Palomar Observatory in November of 1993. As telescopes were trained on the surprise visitor, the object was seen to be an unusual alignment of luminous dots, 'strung out like pearls on a string' (see figure 8.2). By working out its trajectory back in time, Brian Marsden and his associates at the Smithsonian Astrophysical Observatory established that this was a comet which had recently broken up in pieces – thus its elongated appearance – after passing too close to planet Jupiter on its previous orbit.

Moreover, the astronomers calculated that the dislocated comet fragments were effectively trapped by Jupiter's gravity, and were swinging back on a collision course with the giant planet for a head-on strike due the following summer. Astronomers calculated with deadly precision that the salvo of impacts would span the whole week of July 16 to July 22, 1994 – due to the length of the cometary train which by then stretched over millions of kilometers.

For the first time in history, a pre-announced comet impact would be directly observable from Earth. As the whole world watched – from the Hubble Space Telescope down to amateur astronomers in their back yards – comet Shoemaker–Levy 9 began its impact demonstration right on schedule, July 16, 1994.

As each fragment plunged into the Jovian atmosphere at velocities in excess of 60 km/s, the display of celestial fireworks ran a full gamut of forms. Each impact typically began with the impingement of a fine dust cloud leading the solid fragment, blasting the atmosphere like a charge of shot and lighting up the gases in a precursory glow. This prelude was followed by a brilliant flash marking the plunge of the rocky fragment itself, which disintegrated in a fireball of superheated plasma billowing 3000 km above the cloud tops. As the impact plume expanded in the vacuum of space, its particles recondensed to rain down upon the Jovian atmosphere

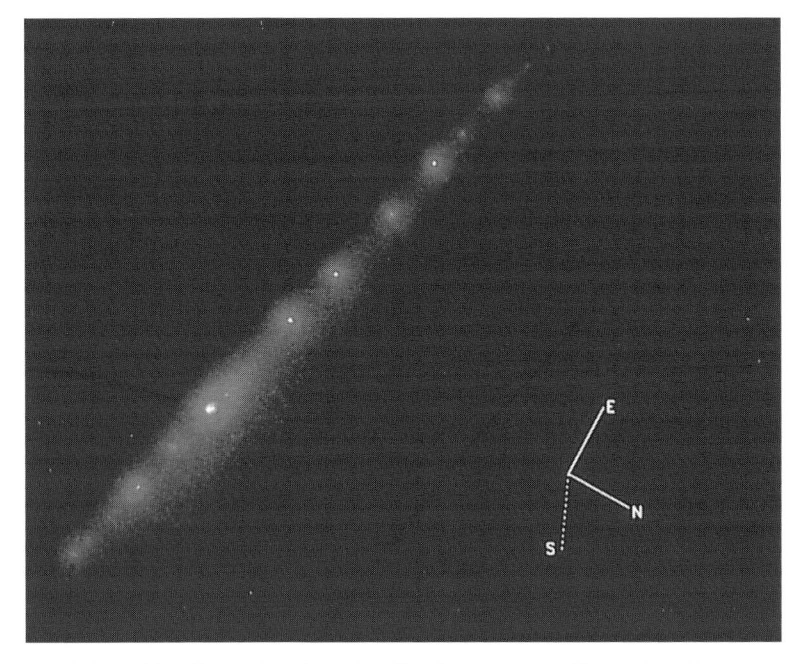

Figure 8.2. The 'string of pearls' of broken-up comet Shoemaker–Levy 9, photographed by the Hubble Space Telescope as it headed for planet Jupiter. The blocks of ice and dust, each measuring up to 500 meters across, collided with Jupiter in a drawn-out salvo that lasted a whole week from July 16 to 22, 1994, blasting the planet with a total impact energy of several million megatons of TNT. (*Photograph: NASA, H. A. Weaver and T. E. Smith, STS, courtesy of ESA.*)

in a secondary shower, visible at infrared wavelengths for several minutes after the main event – the very 'roasting' effect that was proposed for the Chicxulub ejecta on Earth. The 'nuclear winter' scene of the K-T scenario was similarly played out. Astronomers were able to observe the spreading of opaque clouds of dust in the Jovian atmosphere, expanding to rings and blotches the size of planet Earth (see figure 8.3). These atmospheric disturbances lasted for months.

When everything was said and done, the sequence of impacts on Jupiter was estimated to total several million megatons of TNT – albeit only a fraction of the energy released in the larger Chicxulub event. With individual impacting fragments measuring no more than 500 m in size, the initial 'whole' comet (before dislocation) measured

Figure 8.3. Fragment 'G' of comet Shoemaker–Levy 9 blasted a wide perturbation in Jupiter's atmosphere as it hit the planet on July 18, 1994. The image was collected by the Hubble Space Telescope, 1 h 45 min after impact, when the rotation of the planet brought the disturbed area into view. The central spot measures 2500 km in diameter, and the exterior crescent 12 000 km – the size of Earth. Despite the sizeable perturbation, the 'G' fragment impact was a thousand times less energetic than the Chicxulub impact on Earth. (*Photograph: NASA, H. Hammel, MIT, courtesy of ESA.*)

only 2 km in diameter. Objects of this size are expected to hit Earth every million years or so, causing regional to global destruction.[8]

The asteroids

With the near miss of asteroid 1989FC (Asclepius) and the real crash of comet Shoemaker–Levy 9 on Jupiter, the world has become

[8] Because of its superior size, Jupiter presents a larger cross section than the Earth to incoming comets and asteroids, and pulls them in from farther out in space. Collisions are therefore about 1000 times more frequent on Jupiter than on Earth: impacts of the kind of Shoemaker–Levy 9 are expected every few centuries on the giant planet.

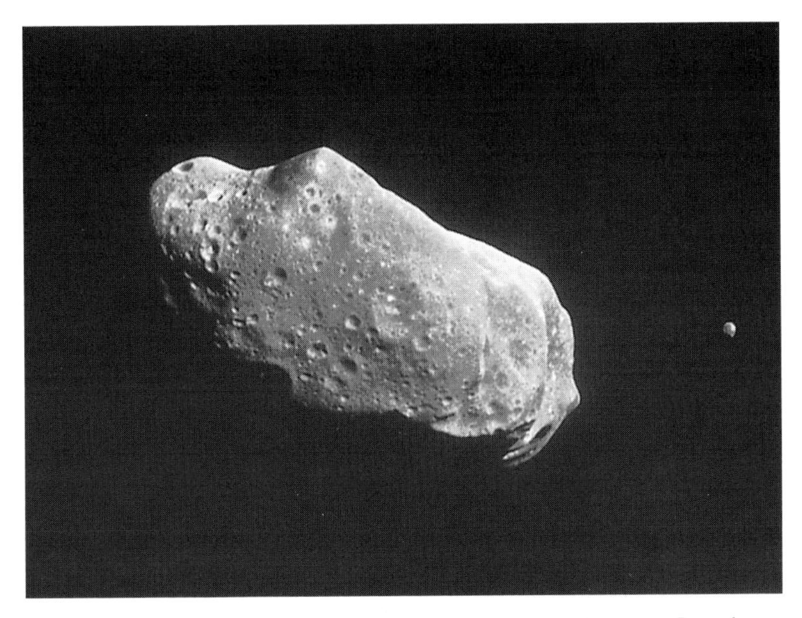

Figure 8.4. Asteroid Ida, 56 km across, and its diminutive moon Dactyl (right), photographed by NASA's Galileo spacecraft. Asteroids are one of the major threats to our planet, since they can be nudged on Earth-crossing trajectories by perturbations due to neighboring planets Jupiter and Mars. (*Photograph: NASA/JPL.*)

brutally aware that the threat of a major impact concerns us here and now, and cannot be pushed back into the distant future. Moreover, we now have the means to detect such threatening objects in space and monitor their trajectories.

It is customary to classify cosmic projectiles into two main families – asteroids and comets – although the distinction is somewhat theoretical (some asteroids are 'burnt out' comets) and the inflicted damage is similar.

Asteroids are solid blocks of rock, metal or dusty ice, and their orbits are mostly contained between those of Mars and Jupiter, at a safe distance from the Earth.

The first and largest asteroid was discovered on January 1, 1801: named Ceres, it is nearly spherical and measures about 1000 km in diameter – a mini-planet of sorts. Only ten asteroids are larger than 250 km. Their numbers shoot up dramatically toward smaller sizes

– the rule of thumb being a one-hundred-fold increase in population for a tenfold decrease in size. Overall, it is estimated that there are hundreds of thousands of asteroids one kilometer in size or larger circling the Sun. Only a very few have been imaged by passing spacecraft (see figure 8.4), but astronomical observations make it possible to classify asteroids in a number of groups, based on their luminosity and color, compared with the spectral properties of known meteorites in the lab.

One dominant group is the 'S' or stony asteroids, which appear to be made up of common silicate minerals like olivine and pyroxene, and are particularly abundant in the inner asteroidal belt (the bodies closest to Mars). Less common are the 'M' or metallic asteroids, made up mostly of iron and nickel. As for the 'C' or carbonaceous asteroids, which appear to be rich in hydrated minerals and carbon-rich molecules, they dominate the outer belt, out to 600 million kilometers from the Sun. Finally, there are a number of minor and stranger classes of asteroids like the clear-colored 'E' or the reddish 'P' and 'D' objects.

Asteroids are believed to be planetary 'building blocks' left over from the early, formative days of the Solar System. They were never able to coalesce into a full-sized planet because of the strong, disturbing influence of nearby Jupiter, preventing the small bodies from merging with one another. On the contrary, Jupiter occasionally ejects drifting asteroids away from their nursery into eccentric orbits that drive them out of the Solar System altogether, or else in the opposite direction, closer in to the Sun. It is these Earth-approaching outcasts that constitute the main impact hazard for our planet.

Earth-crossing asteroids

The first few hundred asteroids discovered by astronomers in the 1800s orbited comfortably far from Earth, but a first surprise came with Eros, spotted in 1898, which approaches within 'only' 20 million kilometers of our planet. In 1932, asteroid Amor was found to swing in even closer (16 million km), soon surpassed by asteroid

Apollo (11 million km). But the real jolt came in 1937 when an asteroid named Hermes was spotted screaming past the Earth at less than a million kilometers, before receding into the darkness of space. It became clear that a number of asteroids came dangerously close to Earth, and that some went so far as to cross the lane of our own orbit in their eccentric race around the Sun.

As astronomers began dedicated searches in the 1970s to look for these near-Earth asteroids, they identified three classes of hazardous bodies. One consists of bodies which swing in from an orbit external to the Earth's and come dangerously close, without actually crossing our path: these are named *Amors*, after the 'flag-ship' body discovered in 1932. Including Amor, Eros, Betulia, Pele and Dionysus, these asteroids are too close for comfort in that they can be perturbed by planet Jupiter or even Mars into more eccentric orbits, that can turn them into real Earth-crossers in the near future.

More dangerous yet are the *Apollos* or true Earth-crossers, which at present swing across our orbit, although they might do so above or below its plane (the *ecliptic*) and thus avoid us on most passes. The problem occurs when the continuous 'twist' of the asteroid's orbit – by astronomical phenomena such as *precession* – brings the asteroid's crossing of our orbital plane to occur exactly at our orbital distance to the Sun, thus threatening us with a direct hit if we happen to be traveling through that point at that time. These Earth-crossers include the large 'flagship' asteroid Apollo itself, as well as Hermes, Geographos, Sisyphus, Toutatis, and many smaller bodies like Asclepius (1989FC).

The list of Earth-crossers would not be complete without citing those asteroids which spend the majority of their orbit *inside* the Earth's ellipse around the Sun rather than outside, and intersect our 'lane' only at their farthermost elongation (at *aphelion*). The first body of this type, asteroid Aten, was discovered in 1976 and gave its name to the family, which also includes asteroids Hathor, Amum and Sekhmet.

Approximately one hundred Earth-crossers were listed by the year 1990. Of those, 65% were Earth-crossing *Apollos*, 10% were Earth-crossing *Atens*, and 25% were Earth-approaching *Amors*. They cover the full range of asteroidal types, the majority being dark, carbon-rich 'C' asteroids, followed by stony and stony-iron 'S' asteroids.

With respect to size, most Earth-crossing asteroids are smaller than 10 km in diameter, except for Ganymede (41 km), Eros (23 km) and Eric (12 km), which are *Amors* anyway and don't at present cross our orbit. Of the true Earth-crossers, Ivar and Betulia are both 8 km in size (virtually as large as the body responsible for the K-T mass extinction), and at least ten other *Apollos* are larger than 5 km, with many more yet to be discovered.

Watching the sky

Today, the search for Earth-crossing asteroids is reaching new heights. Thanks to the pioneering efforts of less than half a dozen dedicated astronomers in the sixties, seventies and eighties, and the public and political awareness of the nineties, the scientific community is standing on the threshold of a rational, upgraded survey of the sky as we roll into the twenty-first century.

Getting this far was an uphill battle. The discovery of Earth-crossers remained a serendipitous matter for years, asteroids showing up by chance on photographic surveys of other celestial bodies – stars, nebulae and distant galaxies. The search for Earth-crossers took on a sharper focus in the 1960s when the Dutch couple Kees and Ingrid van Houten and Dutch-born astronomer Tom Gehrels, working at the Palomar and Leiden observatories, found four near-Earth asteroids in 1960 alone.

In 1972, Eleanor Helin and Eugene Shoemaker joined in the hunt and started the Planet-Crossing Asteroid Survey at Palomar Mountain in the 1980s, and also introduced a new photographic technique to detect fast-moving asteroids. Helin doubled the count of identified near-Earth asteroids from a total of 20 to 40.

Eugene Shoemaker started a twin survey program in 1982 with his wife Carolyn and, joined by David Levy in the 1990s, extended their search to approaching comets (one of their claims to fame would be the Jupiter-bound comet Shoemaker–Levy 9).

The search gathered momentum in 1989 when Tom Gehrels at

Figure 8.5. The Spacewatch telescopes atop Kitt Peak, Arizona, are dedicated to the search of near-Earth asteroids and comets. Manned by Tom Gehrels and a team of dedicated astronomers, the telescopes detect several new objects every night. (*Photograph: by the author.*)

the University of Arizona put together a dedicated search program on Kitt Peak, near Tucson. Named Spacewatch, it makes use of a CCD technique rather than photographic plates to detect nearby asteroids. Manned by Gehrels, Jim Scotti, David Rabinowitz, Robert Jedicke, Robert McMillen, Marcus Perry and others, the dedicated search is now turning out nearly half of all near-Earth discoveries, and the team has now added a 1.8-m instrument – four times more sensitive than the 0.9-m telescope – to their program (see figure 8.5).

The University of Arizona team at Kitt Peak carried the torch virtually alone in the mid-1990s, while other programs were only beginning to get off the ground. Among these, Ted Bowell is setting up a new search program at Lowell Observatory in Flagstaff using CCDs; in 1995 Eleanor Helin started a new program on the island of Maui with the U.S. Air Force (NEAT); and a new program (LINEAR) started up in New Mexico in 1997. Amateur astronomers in Japan are also contributing to the search, as might China in the near future.

There have also been some disappointing shutdowns. Duncan Steel ran a photographic detection program for years at the Anglo-Australian Observatory with Rob McNaught and Ken Russell, until they were closed down by the Australian government on January 1, 1997. Alain Maury of the Côte d'Azur Observatory in France has likewise received little support in his attempt to run a search program in Europe with Gerhard Hahn of the German Space Agency.

Searching for asteroids is a special exercise that demands specific telescopes and detection procedures. Nearby asteroids are observed as faint specks of light moving across the starry background. One way to detect them is to take pairs of images of a region of the sky, an hour apart, and check for celestial objects that move from one frame to the next. Once a moving asteroid is identified, its orbit is calculated from its change in position, generally through the services of Brian Marsden and his two-person team at the Smithsonian's Astrophysical Observatory in Massachusetts.

Thanks to the combined efforts of the Palomar surveys and the Spacewatch effort at Kitt Peak, the number of detected near-Earth asteroids has surged from 100 in 1990 to over 400 in 1998, the discovery rate averaging close to 50 new bodies per year. Half of them are over one kilometer in size, and the other half consists of smaller objects ranging from hundreds of meters down to a few meters in size.

Results are pouring in, but the truth of the matter is that the search is still in its infancy: it is expected that there are about 2000 near-Earth asteroids larger than one kilometer. Approximately 200 have been found as of 1998, which is 10% of the suspected total. As for the smaller near-Earth objects, their census will also grow in the course of the sky survey: it is estimated that there are 10 000 bodies down to 500 meters in size. Clearly, there is still much work to be done.

A sense of urgency

If we are to tackle the problem responsibly and protect our civilization from the next destructive impact, we need to speed up and coordinate the near-Earth sky survey. In the past, astronomers have failed to convince lawmakers and funding agencies of the necessity and urgency of such a survey, but with the close passage of asteroid 1989FC and the Jupiter comet crash of 1994, the mood is beginning to change.

The 1989 asteroid alert was enough to spark concern in the U.S. Congress: two workshops were organized to define both a detection strategy (chaired by David Morrison of NASA's Ames Research Center) and an interception strategy (chaired by Jurgen Rahe and John Rather, also of NASA). As the Detection Committee worked on its report for the U.S. Congress (which it presented in March of 1993), other reflection groups were organized around the world: the International Astronomical Union put together a workshop in 1991, and in the wake of an international seminar at Vulcano Island, Sicily, established the Spaceguard Foundation, pooling the thoughts and resources of all experts. The Council of Europe issued a resolution promoting 'the establishment and development of a Space survey foundation . . . to run up an inventory of near-Earth objects as complete as possible', although concrete measures have still to be taken.

The threat of asteroid impacts is now out in the open and debated in both scientific and political circles. But financial backing is still scarce, despite the growing awareness of the impact hazard: as of 1998, search programs are still proceeding on shoestring budgets, employing less than two dozen people worldwide.

Reflection groups, such as the U.S.-sponsored Detection Committee, have suggested that a centralized, international search program should be implemented and funded at the level of several million U.S. dollars per year. Advocates of this Spaceguard project suggest that six to ten dedicated telescopes, with apertures of up to two meters, should be specially built and installed in both hemispheres to tackle the task. But even if such a centralized project were slated and funded, it would take close to 15 years to install it – an interval during which some of the technical and strategic choices of the system might turn out to be flawed or outdated. As Tom Gehrels has suggested, it might be wiser to spend the time and money on already existing programs. For the issue is not so much a lack of funding, or the need for a giant, expensive system, but one of commitment, on a day-to-day basis: more astronomers must devote their time and energy to doing the painstaking work needed to detect near-Earth asteroids. During the second semester of 1998, in fact, there has been some improvement. NASA has doubled its asteroid-hunting budget to three million dollars annually, and the discovery rate of near-Earth bodies over a kilometer in size is surging to about 50 new objects per year.

The threat of comets

Asteroids represent only one side of the issue. Another component of the impact hazard comes from the eccentric and much less predictable Earth-crossing comets.

Asteroids circle the Sun in relatively circular orbits, close to the Earth's orbital plane. Comets are much more eccentric and travel on elongated orbits that carry them inward from the outer reaches of the Solar System, past the orbits of the giant planets – which perturb their trajectories – and occasionally across our own orbital path as they swing in close to the Sun.

Most comets are believed to reside billions of kilometers from Earth in a halo of 'dirty snowballs' known as the Oort cloud. Gravitational perturbations – such as the passing of a nearby star or a giant molecular cloud – sporadically propel a number of these cometary bodies toward the Sun where they cross the path of the giant planets and occasionally end up trapped in shorter, Earth-crossing orbits.

It is customary to classify comets into three groups, according to their orbital status: long-period comets, which take hundreds to thousands of years to complete a loop around the Sun; intermediate-period comets, which have been reined in by the giant planets and complete an orbit in less than 200 years; and short-period comets whose revolution takes less than 20 years. About 10% of the latter are Earth-crossing, as are a number of the intermediate- and long-period objects.

Some comets are very large, such as the exceptional Sarabat comet observed in 1729, which was probably 200 to 300 km across, but most comets are under 20 km in size (Halley's nucleus probably

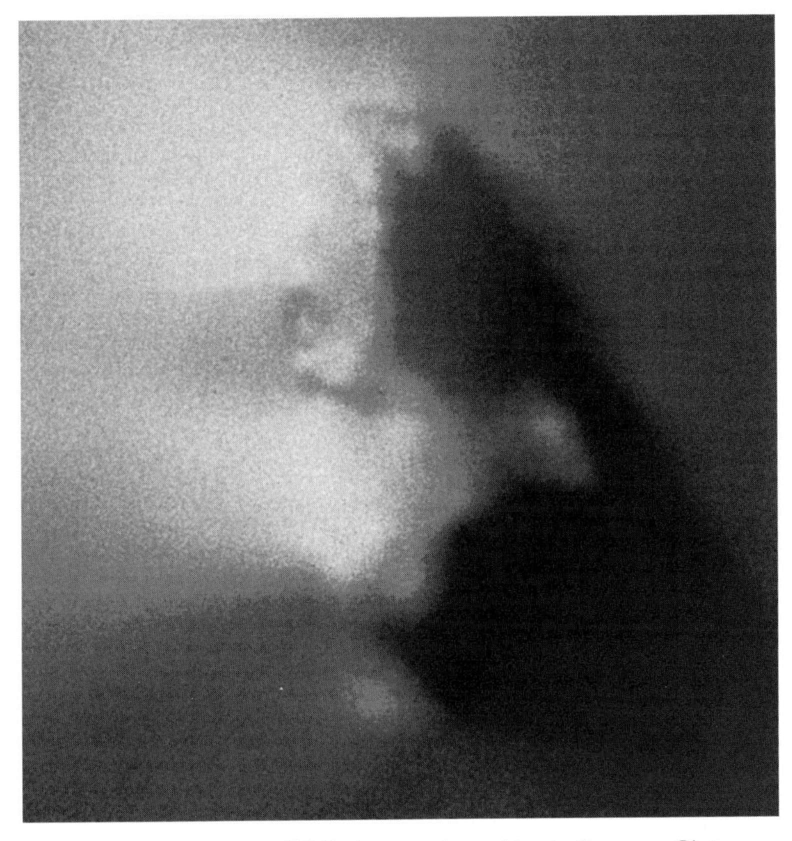

Figure 8.6. The core of Halley's comet, imaged by the European Giotto space probe in 1986. The irregular-shaped central body measures 16 by 8 km. The whitish shroud on its left side is caused by the sublimation of ices heated by sunlight. The jets of gas act as reactors and constantly modify the comet's trajectory. Comets are less predictable than asteroids and constitute a major threat for life on Earth. (*Photograph: ESA/Max Planck Institut für Aeronomie, H. U. Keller.*)

measures 20 by 10 km). Comets are believed to be less dense than asteroids: because of their genesis in the cold, outer reaches of the Solar System, they are made up of low density, frozen volatiles like water. On the other hand, comets are often faster moving than asteroids: whereas the latter cross the Earth's orbit at typical velocities of 20 to 40 km/s, Earth-crossing comets on head-on trajectories reach relative speeds as high as 72 km/s.

Over the last few years, the study of comets and asteroids has somewhat blurred the distinction between both classes of objects. It appears that, over the aeons, the asteroid belt should have dwindled to a much smaller population than observed at present, through removal by planetary impacts, were it not replenished by new objects coming in from outer space – i.e. captured comets. A number of asteroids are therefore believed to be 'extinct' comets or cometary fragments, no longer capable of sustaining a shiny, volatile-rich appearance.

This periodic replenishment of the Earth-crossing population leads to the notion of *coherent catastrophism.* Coined by Australian astronomer Duncan Steel, the concept of coherent catastrophism arises from the fact that, episodically, a large comet might come rushing into the inner Solar System to get trapped there by the gravitational tug of the giant planets. If the deflection stress is brutal, the loosely bound body of ice and dust might break up into fragments – the sort of stress that dislocated comet Shoemaker–Levy 9 before it impacted Jupiter.[9]

Could the risk of impacts on Earth surge each time a giant comet breaks up into a large number of projectiles, that would then cross the Earth's orbit in a coherent stream? Duncan Steel believes that this scenario would account for the clusters of impacts surmised in the geological and historical records. Computer simulations suggest that a giant comet might get trapped in a short-period orbit (and likely disintegrate) every one hundred thousand years on average. After such a break-up, the swarm of debris would periodically cross the Earth's orbit – perhaps once every few millennia, each time showering the planet with an impact pulse lasting decades to centuries.

According to British astronomers Victor Clube and Bill Napier, such an impact pulse took place 3000 to 6000 years ago, when the Earth allegedly suffered a severe deterioration of the climate (with major flooding in Mesopotamia and Egypt) and saw the rise of astronomy – perhaps as a catastrophe-predicting science – in the burgeoning civilizations of Egypt, South America and England.

[9] Comet break-ups are a frequent occurrence: they are signaled in ancient Chinese records, and recent examples include the splitting of comet Biela in 1846; comet C/West in 1975; and comet 73P/Schwassmann–Wachmann-3 in 1995.

This view is also championed by Duncan Steel, who calls attention to the unusual density of cometary dust in the inner Solar System (as testified by the zodiacal light that glows in the sky right after sunset or before dawn). Perhaps a substantial number of the short-period comets now circling the Sun are remnants of such a break-up.

One particularly intriguing body in this respect is comet Encke, a 5-km object which circles the Sun in 3.3 years. Each time we cross its orbital path (twice a year, in late June and early November), we are struck by a shower of shooting stars – small dust particles, spread along the comet's path. These are known as the Taurid (November) and Beta-Taurid (June) meteor showers. Occasionally, larger fragments of the comet are known to strike. Such is apparently the case of the Tunguska event, which took place on the 30th of June, at the peak of the Beta-Taurid shower. The bolide originated from the proper direction in the sky – the Taurus constellation – to be an Encke comet fragment.

Also connected to the Encke swarm are the events of June 1975, when seismometers installed on the Moon by the American astronauts registered a flurry of impacts on our satellite's near side, the largest projectiles weighing close to a ton.

Equally compelling is the fact that a number of Earth-crossing asteroids have orbits similar to Encke's comet and might be other fragments originating from the break-up of a larger, pre-Encke body.

When did this cometary break-up take place, setting up the postulated, recent surge of impacts in the Earth–Moon system? An examination of the microscopic craters that pepper the lunar rocks hints that there was a greatly increased flux of micrometeorites hitting the Moon about 20 000 years ago. Ice cores from the Arctic also point to increased concentrations of meteoritic dust on Earth at that time. Moreover, if we plot the orbit of the Taurid meteor stream (and of a few other meteor streams) back in time, we find that they converge towards a common origin – likely the parent comet at the time of break-up – some 20 000 years ago. The onset of a coherent surge of impacts at that time might also explain the climatic upheaval that ended the ice age and prompted the rise of civilization.

Be it as it may, comets and their broken-up fragments constitute a sizeable percentage of the impact hazard on Earth (estimates range from 2% to 25% of 'hits'). Those which have been trapped into short-period orbits are little different from asteroids, and their detection thus requires the same basic techniques as for Earth-crossing asteroids.

On the other hand, those 'new' comets which sporadically enter the inner Solar System from afar pose a distinct problem. Their surprise appearances require detection as far ahead of time as possible, in order to calculate their orbits and pinpoint their chances of impact with a reasonable lead time for mankind's reaction and protection. Moreover, comets can come from any direction in the sky, as opposed to asteroids, which are mostly clustered around the ecliptic (the 'equatorial' plane of the Earth's course around the Sun). To spot these comets, dedicated telescopes are needed to peer farther out into space, not only along the ecliptic but out to the celestial poles as well. The Spacewatch 1.8 meter telescope at Kitt Peak is one such instrument.

Protecting the Earth

The census of Earth-crossing asteroids and comets is reassuring so far. Calculations show that none of the large objects tracked at present are bound to collide with our planet in the foreseeable future. Asteroid Toutatis (5 km) will perform the closest pass in the first quarter of the twenty-first century, speeding by Earth at a respectable 1.5 million kilometers on September 29, 2004; and the closest call by a known comet will be by the unpronounceable Honda-Mrkos-Pajdusakova on August 15, 2011, at a comfortable distance of 9 million kilometers. Apart from new and unpredictable comets flying in from the outer Solar System, there doesn't seem to be any palpable threat of a global mass extinction by a kilometer-sized object within the next few decades.[10]

[10] Projections into the more distant future are unreliable because of the cumulation of errors and uncertainties over many orbits (concerning both the asteroids and the perturbing planets). It is worth noting, however, that computer runs by Paolo Farinella at the University of Pisa show a slight chance of collision with asteroid Eros (20 km in size), approximately one million years into the future.

The hazard is much less constrained when it comes to smaller objects under a kilometer in size (which could still destroy entire regions or cities). Most of these objects are not yet on record, and are usually spotted at the last minute: on May 19, 1996, for example, an asteroid over 300 meters in size sped by Earth at roughly the distance of the Moon (450 000 km), and it was signaled only a few *days* before its 'grazing' pass. Clearly, these numerous small objects present the greatest short-term threat to our planet.[11]

Beyond the problem of detection lies the question of response. What could we possibly do, if we did predict a cosmic hit days or months ahead of time?

Besides the population taking cover (in underground shelters, for example) and bracing itself for the impact, a number of preventive measures are imaginable. Science fiction writers have mulled over the problem for years, and scientists first began to tackle the issue in June of 1968, when the 'close' passage of asteroid Icarus (seven million kilometers from Earth) prompted MIT professors to stage a theoretical exercise for their students. The answer to the threat was simple enough and would set the basis for all future studies: with its newly developed heavy launchers and atomic weapons, mankind has the capability to intercept an incoming asteroid or comet and supply an explosive impulse large enough to deflect it off its Earth-bound trajectory. Gentle deflection is preferable to blowing the body to pieces: breaking up an asteroid or comet would simply transform one lone impactor into a salvo of smaller projectiles, threatening the Earth with multiple impacts.

How much thrust (i.e. nuclear power) would it take to divert an Earth-crosser from its collision course? In fact, the problem boils down to a time issue – the amount of warning time available to perform the interception. The deflection impulse must be initiated as far away from Earth as possible, so that it can progressively steer the threatening body away from its trajectory over time. Spacecraft navigators know the maneuvre well: a small change in velocity,

[11] There has been one recent 'false alert'. A one-kilometer sized asteroid, discovered in 1997, was calculated in March 1998 to be headed on a potential collision course with Earth during one of its future passes, in October of 2028. The press picked up the story, but further calculations showed that the asteroid will miss the Earth that year by close to a million kilometers.

Figure 8.7. The exploration of asteroids and comets by space probes constitutes the first step towards designing an efficient program of deflection, should they one day threaten the Earth. Only by measuring the texture, structure and center of gravity of such bodies will it be possible to correctly model the effects of stand-off explosions and other 'nudges' on their trajectory. (*Photograph: JPL/Ken Hodges.*)

on the order of ten centimeters per second, will deflect a mobile one thousand kilometers in two months time, and six thousand kilometers – the radius of target Earth – in one year. Hence, if a 10 cm/s impulse is communicated to a threatening asteroid one year before collision time, it is sufficient to steer it off course.[12] Conversely, if a threatening object is detected and reached less than a year before impact, a higher deflecting impulse must be provided to steer it away in due time.

Once the detection of a threatening body is accomplished with as much lead time as possible, the interception is fairly straightfor-

[12] In fact this is a minimum figure. Because of the error margin in the calculation of the object's trajectory, one would want to confer a larger deflection to a threatening object than called for in theory, to make sure it really misses the Earth.

ward. One would launch an interplanetary probe carrying the right amount of nuclear explosives to provide the necessary deflective thrust. Calculations show that one megaton of TNT (one of our medium-sized nuclear weapons) would provide a velocity change of 10 cm/s to an asteroid one kilometer in size, if the explosive energy were perfectly transformed into useful acceleration. Considering heat loss and other forms of energy degradation, a nuclear bomb several times more powerful than the theoretical figure would be needed, but such bombs do exist, as well as the heavy boosters required for the launch (American Titan III and IV; Russian Proton and Energya rockets).

There is some discussion as to the technique of detonation, in order to obtain maximum coupling of the energy with the asteroid or comet. The first solution that comes to mind is to drive the nuclear bomb into the surface of the target before triggering the explosion, but this solution carries the risk of fragmenting the body and ending up with an aggravated situation of several projectiles racing on a collision course with Earth, rather than one. Moreover, some of the energy might be dissipated into wasteful rotation of the object rather than linear acceleration, especially if its center of gravity and moment of inertia were badly known.

In workshops set up over the years to discuss this interception issue,[13] the strategy preferred by specialists (such as Thomas Ahrens and Alan Harris of Caltech) is to explode the nuclear charge several hundreds of meters above the target body, evenly spreading the 'nudging' energy over a wide area. The radiation bouncing off such a large area would insure an evenly distributed propulsion force, with little risk of fragmenting the body. Moreover, the explosive yield required for proper deflection appears to be no greater than in the case of a surface-implanted bomb.

Whatever the technique employed, a progressive, multi-step approach might be preferable to a one-off blast. One large explosion would be difficult to carry out with the required precision, whereas

[13] The 'Interception Committee' exploring the issue of asteroid deflection met at Los Alamos, New Mexico, in January of 1992. The 'think tank' consisted of 90 specialists, all American except for Australian astronomer Duncan Steel. The most comprehensive book on the issue is *Hazards due to Comets and Asteroids*, edited by Tom Gehrels and published by the University of Arizona Press, Tucson, 1994.

several smaller blasts by an armada of interceptors strung out over days or weeks would allow operators back on Earth to analyze each blast to better understand the physical properties of the target body and its response, in order to set up the next explosion and reach, step by step, the required change in trajectory.

Other deflection strategies are proposed as substitutes to nuclear bombs, although not nearly as simple, nor currently operational.[14]

One major conclusion emerges from the brainstorming of the Interception Committee and other workshops: we must learn more about the nature and dynamics of asteroids and comets if we are to properly model deflection scenarios.

Fortunately, a number of space missions are designed to do just that. Launched in February 1997, NASA's NEAR space probe (Near-Earth Asteroid Rendezvous) has flown by asteroid Mathilde on June 27, 1997, taking 500 photos of the carbon-black body. It is now headed for asteroid Eros, which it reached in January 1999. The probe will enter a grazing orbit around the 40-km body and determine its precise shape, mass and gravity field. In July 1999, another NASA spacecraft – the new generation probe Deep Space 1 – will fly by a smaller asteroid (known as 1992 KD), before heading for comet Borrelly, which it will encounter in 2001. Last but not least, Japan's Muses-C will lift off in January 2002 for a rendezvous and touchdown on asteroid Nereus in September 2003 to collect 'soil' samples. After completing its task, Muses-C will lift off on an Earth-bound trajectory and deliver its precious cargo to our home planet by parachute in January 2006.

Comets will also get a thorough review in the first decade of the twenty-first century. NASA's Stardust spacecraft lifted off in February 1999 on a trajectory to encounter comet Wild-2 in January 2004. During a flyby as close as 150 km from the comet's nucleus, the spacecraft will capture cometary dust grains with a 'sticky' gel attached to its panels. The hoist of comet dust will be stored aboard a reentry capsule and delivered by parachute in January 2006, only

[14] One such technique is the solar sail, by which giant riggings of reflective material would be hoisted on asteroids and comets, bouncing off the solar wind and steering the celestial bodies in a neatly controlled regatta. However attractive, the sailing solution meets with a number of difficulties, including the fact that most comets and asteroids spin and tumble along their orbits, and stabilizing them to properly use the sail would prove too formidable a task.

days before Japan's Muses-C delivers its own cargo of asteroidal material.

Meanwhile, NASA's CONTOUR spacecraft will encounter at a safe distance no less than three comets between 2003 and 2008: comets Encke, Schwassmann–Wachmann-3 and d'Arrest. Finally, the last two comet missions of the decade will be the most ambitious : NASA's Deep Space 4 and ESA's Rosetta.

Deep Space 4 will depart in April 2003 for an encounter with comet Tempel-1 in late 2005. It will spend several months in orbit and dispatch a lander on the surface that will drill one meter into the ground and collect samples. These will be returned to Earth by a reentry vehicle in June 2010.

Also to be launched in 2003, the European-built Rosetta will fly by two asteroids before going in orbit around comet Wirtanen for three years, watching it spew out volatiles as it approaches the Sun, and landing a probe on its surface in August 2012. The lander will anchor itself to the comet's surface with a harpoon, drill two meters into the ground and radio back a stream of imagery and chemical data.

Thus, as we enter the third millennium, the detection of Earth-crossing asteroids and comets will be in full swing, as well as the study of their physical and dynamic properties. Strategies for the interception of threatening bodies will be refined, although there is no urgent need at this time to prepare missiles or place bombs in orbit. Indeed, as Carl Sagan and others have pointed out, there is some risk that if such weapons are deployed, they could be misused to stage a malevolent strike or a terrorist attack aimed back at Earth.

Despite this risk, it is remarkable that mankind has found a peaceful – and even vital – use for its nuclear weaponry, instead of using it for self-destruction. Perhaps the saving grace of our civilization is to reach awareness of the impact threat and develop the technical means to neutralize it, before a catastrophic hit has the chance to wipe us off the planet and reset the game of evolution. In this respect, we can be thankful to Earth scientists for having been curious and clever enough to ponder over, and resolve, the mysterious demise of the dinosaurs.

Bibliography

Books and Compilations

Alvarez, W., *T. rex and the Crater of Doom*, Princeton University Press, Princeton, 1997.

Buffetaut, E., *Dinosaures: à la Recherche d'un Monde Perdu*, L'Archipel, Paris, 1997.

Chapman, C.R. and Morrison, D., *Cosmic Catastrophes*, Plenum Press, New York, 1989.

Fastovsky, D.E. and Weishampel, D.B., *The Evolution and Extinction of the Dinosaurs*, Cambridge University Press, Cambridge, 1996.

Gehrels, T., editor, *Hazards due to Comets and Asteroids*, University of Arizona Press, Tucson, 1994.

Glenn, W., editor, *The Mass Extinction Debates: How Science Works in a Crisis*, Stanford University Press, Stanford, 1994.

Hodge, P., *Meteorite Craters and Impact Structures of the Earth*, Cambridge University Press, Cambridge, 1994.

Hsü, K., *The Great Dying: Cosmic Catastrophe*, Harcourt Brace Jovanovich, New York, 1986.

Lewis, J.S., *Rain of Iron and Ice*, Helix Books, Addison-Wesley Publishing Company, 1996.

Melosh, H.J., *Impact Cratering: A Geological Process*, Oxford University Press, 1989.

Officer, C. and Page, J., *The Great Dinosaur Extinction Controversy*, Helix Books, Addison-Wesley Publishing Company, 1996.

Raup, D.M., *Extinction: Bad Genes or Bad Luck?*, W.W. Norton & Company, London & New York, 1991.

Russell, D.A., *The Dinosaurs of North America*, University of Toronto Press, Toronto, 1989.

Ryder, G., Fastovsky, D. and Gartner, S., editors, *The Cretaceous–Tertiary Event and Other Catastrophes in Earth History*, Geological Society of America SP-307, Boulder, 1996.

Sharpton, V.L. and Ward, P.D., editors, *Global Catastrophes in Earth History*, Geological Society of America SP-247, Boulder, 1990.

Shaw, H., *Craters, Cosmos and Chronicles: a New Theory of Earth*, Stanford University Press, Stanford, 1994.

Steel, D., *Rogue Asteroids and Doomsday Comets*, John Wiley & Sons, New York, 1995.

Verschuur, G.L., *Impact! The Threat of Comets and Asteroids*, Oxford University Press, Oxford and New York, 1996.

Articles

Chapter 1

Alvarez, W., Alvarez, L.W., Asaro, E. and Michel, H.V., The end of the Cretaceous: sharp boundary or gradual transition?, *Science*, **223**, 1183–86, 1984.

Buffetaut, E., Cuny, G. and Le Lœuff, J., French dinosaurs: the best record in Europe?, *Modern Geology*, **16**, 17–42, 1991.

Raup, D.M. and Sepkoski, J.J., Mass extinctions in the marine fossil record, *Science*, **215**, 1501–3, 1982.

Russell, D.A., The enigma of the extinction of the dinosaurs, *Ann. Rev. Earth Planet. Sci.*, **7**, 163–82, 1979.

Sheehan, P.M., Fastovsky, D.E., Hoffmann, R.G., Berghaus, C.B. and Gabriel, D.L., Sudden extinction of the dinosaurs: Latest Cretaceous, Upper Great Plains, USA, *Science*, **254**, 835–9, 1991.

Wolfe, J.A. and Upchurch, G.R., Vegetation, climatic and floral changes at the Cretaceous–Tertiary boundary, *Nature*, **324**, 148–52, 1986.

Chapter 2

Alvarez, L.W., Alvarez, W., Asaro, F. and Michel, H.V., Extraterrestrial cause for the Cretaceous–Tertiary extinction, *Science*, **208**, 1095–1108, 1980.

Bohor, F.B., Foord, E.E., Modreski, P.J. and Triplehorn, D.M., Mineralogic evidence for an impact event at the Cretaceous–Tertiary boundary, *Science*, **224**, 867–9, 1984.

Carlisle, D.B., Diamonds at the K/T boundary, *Nature*, **357**, 119–20, 1992.

Ganapathy, R., A major meteorite impact on the Earth 65 million years ago: Evidence from the Cretaceous–Tertiary boundary clay, *Science*, **209**, 921–3, 1980.

Hsü, K.J., Terrestrial catastrophe caused by cometary impact at the end of Cretaceous, *Nature*, **285**, 201–3, 1980.

Robin, E., Bronté, P., Froget, L., Jéhanno, C. and Rocchia, R., Formation of spinels in cosmic objects during atmospheric entry: a clue to the Cretaceous/Tertiary boundary event, *Earth Planet. Sci. Lett.*, **108**, 181–90, 1992.

Smit, J. and Hertogen, J., An extraterrestrial event at the Cretaceous–Tertiary boundary, *Nature*, **285**, 198–200, 1980.

Smit, J. and Klaver, G., Sanidine spherules at the Cretaceous–Tertiary boundary indicate a large impact event, *Nature*, **292**, 47–9, 1981.

Chapter 3

Bhandari, N., Shukla, P.N., Ghevariya, Z.G. and Sundaram, S.M., Impact did not trigger Deccan volcanism: evidence from Anjar K/T boundary intertrappean sediments, *Geophys. Res. Lett.*, **22**, 433–6, 1995.

Carter, N.L., Officer, C.B., Chesner, C.A. and Rose, W.I., Dynamic deformation of volcanic ejecta from the Toba caldera: possible relevance to Cretaceous/Tertiary boundary phenomena, *Geology*, **14**, 380–3, 1986.

Courtillot, V., Besse, J., Vandamme, D., Montigny, R., Jaeger, J.-J. and Cappetta, H., Deccan flood basalts at the Cretaceous/Tertiary boundary?, *Earth Planet. Sci. Lett.*, **80**, 361–74, 1986.

Hallam, A., End-Cretaceous mass extinction event: argument for terrestrial causation, *Science*, **238**, 1237–42, 1986.

Hoffman, A. and Nitecki, M.H., Reception of the asteroid hypothesis of terminal Cretaceous extinctions, *Geology*, **13**, 884–7, 1985.

Keller, G., Analysis of El Kef blind test I, *Marine Micropaleontology*, **29**, 65–103, 1997.

Marshall, C.R. and Ward, P.D., Sudden and gradual molluscan extinctions in the Latest Cretaceous of Western European Tethys, *Science*, **274**, 1360–3, 1996.

Officer, C.B. and Drake, C.L., Terminal Cretaceous environmental events, *Science*, **227**, 1161–7, 1985.

Smit, J. and Nederbragt, A.J., Analysis of the El Kef blind test II, *Marine Micropaleontology*, **29**, 65–103, 1997.

Stinnesbeck, W., Barbarin, J.M., Keller, G. *et al.*, Deposition of channel deposits near the Cretaceous–Tertiary boundary in northeastern

Mexico: catastrophic or 'normal' sedimentary deposits?, *Geology*, **21**, 797–800, 1993.

Venkatesan, T.R., Pande, K. and Gopalan, K., Did Deccan volcanism pre-date the Cretaceous–Tertiary transition?, *Earth Planet. Sci. Lett.*, **119**, 181–9, 1993.

Chapter 4

French, B.M., 25 years of the impact-volcanic controversy: is there anything new under the Sun or inside the Earth?, *EOS*, **71**, 411–14, 1990.

Grieve, R.A.F., Terrestrial impact structures, *Ann. Rev. Earth Planet. Sci.*, **15**, 245–70, 1987.

Grieve, R.A.F., Terrestrial impact: the record in the rocks, *Meteoritics*, **26**, 175–94, 1991.

Grieve, R.A.F., Stöffler, D. and Deutsch, A., The Sudbury structure: controversial or misunderstood?, *J. Geophys. Res.*, **96**, 22753–64, 1991.

Izett, G.A., Cobban, W.A., Obradovich, J.D. and Kunk, M.J., The Manson impact structure: $^{40}Ar/^{39}Ar$ age and its distal impact ejecta in the Pierre Shale in southeastern South Dakota, *Science*, **262**, 729–32, 1993.

Jansa, L.F. and Pe-Piper, G., Identification of an underwater extraterrestrial impact crater, *Nature*, **327**, 612–14, 1987.

Koeberl, C., Sharpton, V.L., Murali, A.V. and Burke, K., The Kara and Ust-Kara impact structures (URSS) and their relevance to the K/T boundary, *Geology*, **18**, 50–63, 1990.

Chapter 5

Alvarez, W., Smit, J., Lowrie, W. *et al.*, Proximal impact deposits at the Cretaceous–Tertiary boundary in the Gulf of Mexico: a restudy of DSDP Leg 77 Sites 536 and 540, *Geology*, **20**, 697–700, 1992.

Blum, J.D., Chamberlain, C.P., Hingston, M.P. *et al.*, Isotopic comparisons of K/T boundary impact glass with melt rock from the Chicxulub and Manson impact structures, *Nature*, **364**, 325–7, 1993.

Bourgeois, J., Hansen, T.A., Wiberg, P.L. and Kauffman, E.G., A tsunami deposit at the Cretaceous–Tertiary boundary in Texas, *Science*, **241**, 567–70, 1988.

Hildebrand, A.R., The Cretaceous/Tertiary boundary impact (or The dinosaurs didn't have a chance), *J. R. Astron. Soc. Canada*, **87**, 77–118, 1993.

Hildebrand, A.R. and Boynton, W.V., Proximal Cretaceous–Tertiary boundary impact deposits in the Caribbean, *Science*, **248**, 843–7, 1990.

Hildebrand, A.R., Penfield, G.T., Kring, D.A. *et al.*, Chicxulub Crater: a possible Cretaceous–Tertiary boundary impact crater on the Yucatan Peninsula, Mexico, *Geology*, **19**, 867–71, 1991.

Hildebrand, A.R., Pilkington, M., Connors, M., Ortiz, C. and Chavez, R., Size and structure of the Chicxulub Crater revealed by horizontal gravity gradients and cenotes, *Nature*, **376**, 415–17, 1995.

Kerr, R., Huge impact tied to mass extinction, *Science*, **257**, 878–80, 1992.

Krogh, T.E., Kamo, S.L. and Bohor, B., U-Pb ages of single shocked zircons linking distal K/T ejecta to the Chicxulub crater, *Nature*, **366**, 731–4, 1993.

Maurrasse, F. J.-M. and Sen, G., Impacts, tsunamis, and the Haitian Cretaceous–Tertiary boundary layer, *Science*, **252**, 1690–3, 1991.

Meyerhoff, A.A., Lyons, J.B. and Officer, C.B., Chicxulub structure: a volcanic sequence of late Cretaceous age, *Geology*, **22**, 3–4, 1994.

Morgan, J., Warner, M., Brittan, J. *et al.*, Size and morphology of the Chicxulub impact crater, *Nature*, **390**, 472–6, 1997.

Pilkington, M., Hildebrand, A.R. and Ortiz-Aleman, C., Gravity and magnetic field modeling and structure of the Chicxulub Crater, Mexico, *J. Geophys. Res.*, **99**, 13 147–62, 1994.

Pope, K.O., Ocampo, A.C. and Duller, C.E., Mexican site for K/T impact crater?, *Nature*, **351**, 105, 1991.

Sharpton, V.L., Burke, K., Camargo-Zanoguera, A. *et al.*, Chicxulub multiring impact basin: size and other characteristics derived from gravity analysis, *Science*, **261**, 1564–7, 1993.

Sharpton, V.L., Dalrymple, G.B., Marin, L.E., Ryder, G., Schuraytz, B.C. and Urrutia-Fucugauchi, J., New links between the Chicxulub impact structure and the Cretaceous–Tertiary boundary, *Nature*, **359**, 819–21, 1992.

Sigurdsson, H., D'Hondt, S., Arthur, M.A. *et al.*, Glass from the Cretaceous–Tertiary boundary in Haiti, *Nature*, **349**, 482–7, 1991.

Smit, J., Montanari, A., Swinburne, N.H.M. *et al.*, Tektite-bearing, deep water clastic unit at the Cretaceous–Tertiary boundary in northeastern Mexico, *Geology*, **20**, 99–103, 1992.

Swisher, C.C. III, Grajales-Nishimura, J.M., Montanari, A. *et al.*, Coeval ^{40}Ar/^{39}Ar ages of 65.0 million years ago from Chicxulub Crater melt rock and Cretaceous–Tertiary boundary tektites, *Science*, **257**, 954–8, 1992.

Chapter 6

Alvarez, W., Claeys, P. and Kieffer, S.W., Emplacement of Cretaceous–Tertiary boundary shocked quartz from Chicxulub Crater, *Science*, **269**, 930–5, 1995.

Hsü, K.J., He, Q., McKenzie, J.A. *et al.*, Mass mortality and its environmental and evolutionary causes, *Science*, **216**, 249–56, 1982.

Melosh, H.J., Schneider, N.M., Zahnle, K.J. and Latham, D., Ignition of global wildfires at the Cretaceous/Tertiary boundary, *Nature*, **343**, 251–4, 1990.

O'Keefe, J.D. and Ahrens, J.T., Impact production of CO_2 by the Cretaceous/Tertiary extinction bolide and the resulting heating of the Earth, *Nature*, **338**, 247–9.

Prinn, R.G. and Fegley, B.F., Bolide impacts, acid rain, and biospheric traumas at the Cretaceous–Tertiary boundary, *Earth Planet. Sci. Lett.*, **83**, 1–15, 1987.

Rampino, M.R. and Volk, T., Mass extinctions, atmospheric sulphur and climatic warming at the K/T boundary, *Nature*, **332**, 63–5, 1988.

Schultz, P.H. and D'Hondt, S., Cretaceous–Tertiary (Chicxulub) impact angle and its consequences, *Geology*, **24**, 963–7, 1996.

Sheehan, P.M. and Hansen, T.A., Detritus feeding as a buffer to extinction at the end of the Cretaceous, *Geology*, **14**, 868–70, 1986.

Toon, O.B., Zahnle, K., Morrison, D., Turco, R. and Covey, C., Environmental perturbations caused by the impacts of asteroids and comets, *Rev. Geophys.*, **35**, 41–78, 1997.

Wolbach, W.S., Gilmour, I., Anders, E., Orth, C.J. and Brooks, R.R., Global fire at the Cretaceous–Tertiary boundary, *Nature*, **334**, 665–9, 1988.

Chapter 7

Alvarez, L.W., Mass extinctions caused by large bolide impacts, *Physics Today*, 24–33, July 1987.

Benton, M.J., Late Triassic extinctions and the origin of the dinosaurs, *Science*, **260**, 769–70, 1993.

Bice D.M., Newton, C.R., McCauley, S., Reiners, P.W. and McRoberts, C.A., Shocked quartz at the Triassic–Jurassic boundary in Italy, *Science*, **255**, 443–6, 1992.

Dypvik, H., Gudlaugsson, S.T., Tsikalas, F. *et al.*, Mjølnir structure: An impact crater in the Barents Sea, *Geology*, **24**, 779–82, 1996.

Gersonde, R., Kyte, F.T., Bleil, U. *et al.*, Geological record and reconstruction of the late Pliocene impact of the Eltanin asteroid in the southern ocean, *Nature*, **390**, 357–63, 1997.

Hut, P., Alvarez, W., Elder, W.P. *et al.*, Comet showers as a cause of mass extinctions, *Nature*, **329**, 118–26, 1987.

Leroux, H., Warme, J.E. and Doukhan, J.-C., Shocked quartz in the Alamo breccia, southern Nevada: Evidence for a Devonian impact event, *Geology*, **23**, 1003–6, 1995.

McLaren, D.J. and Goodfellow, W.D., Geological and biological consequences of giant impacts, *Ann. Rev. Earth Planet. Sci.*, **18**, 123–71, 1990.

Poag, S.W., Powars, S.W., Poppe, L.J. *et al.*, DSDP Site 612 bolide event: new evidence of a late Eocene impact-wave deposit and a possible impact site, *Geology*, **20**, 771–4, 1992.

Rampino, M.R. and Haggerty, B.M., Extraterrestrial impacts and mass extinctions, in: *Hazards due to comets and asteroids*, edited by T. Gehrels, University of Arizona Press, Tucson, 1995.

Rampino, M.R. and Haggerty, B.M., The 'Shiva hypothesis': impacts, mass extinctions, and the Galaxy, *Earth, Moon and Planets*, **72**, 441–60, 1996.

Rampino, M.R. and Stothers, R.B., Flood basalt volcanism during the past 250 million years, *Science*, **241**, 663–7, 1988.

Renne, P.R., Zichao, Z., Richards, M.A., Black, M.T. and Basu, A.R., Synchronicity and causal relations between Permian–Triassic boundary crises and Siberian flood volcanism, *Science*, **269**, 1413–16, 1995.

Spray, J.G., Kelley, S.P. and Rowley, D.B., Evidence for a late Triassic multiple impact on Earth, *Nature*, **392**, 171–3, 1998.

Toon, O.B., Zahnle, K., Turco, R.P. and Covey, C., in: *Hazards due to Comets and Asteroids*, edited by T. Gehrels, pp. 791–826, University of Arizona Press, Tucson, 1994.

Chapter 8

Ahrens, T.J. and Harris, A.W., Deflection and fragmentation of near-Earth asteroids, *Nature*, **360**, 429–33, 1992.

Brown, G.E., chairman, *The threat of large earth-orbit crossing asteroids*, Hearing before the Subcommittee on Space, U.S. House of Representatives, 24/3/1993, Washington D.C., ISBN 0–16–040967–5.

Chapman, C.R. and Morrison, D., Impacts on the earth by asteroids and comets: assessing the hazard, *Nature*, **367**, 33–4, 1994.

Chyba, C.F., Thomas, P.J. and Zahnle, K., The 1908 Tunguska explosion: atmospheric detonation of a stony asteroid, *Nature*, **361**, 40–4, 1993.

Gehrels, T., Scanning with charge-coupled devices, *Space Sci. Rev.*, **58**, 347–75, 1991.

Helin, E.F. and Shoemaker, E.M., Palomar planet-crossing asteroid survey 1973–1978, *Icarus*, **40**, 321–8, 1979.

Shoemaker, E.M., Asteroid and comet bombardment of the earth, *Ann. Rev. Earth Planet. Sci.*, **11**, 461–94, 1983.

Index

Page numbers in italic indicate figure captions or footnotes